CW00796996

Seventy
Years
of Farm
Machinery
Part One:
Seed time

TRACK MARSHALL 70
XDX
481L

Part One:
Seed time

Brian Bell MBE

OLD POND PUBLISHING
IPSWICH

First published 2009

ISBN 978-1-906853-18-1

A catalogue record for this book is available from the British Library

Published by
Old Pond Publishing Ltd
Dencora Business Centre, 36 White House Road
Ipswich IP1 5LT United Kingdom

www.oldpond.com

Cover design and book layout by Liz Whatling
Printed and bound in China

Contents

Conversion Table

The information below is offered to help those readers who may be too young to remember the imperial days of pounds, shillings and pence.

1 gallon	=	4.6 litres
1 inch	=	25.4 mm
1 foot	=	300 mm
1 yard	=	910 mm
1 acre	=	0.4 hectare
1cwt	=	50.8 kg
1 ton	=	1,016 kg
1 shilling	=	5 pence
£1	=	20 shillings

Introduction

The twentieth century undoubtedly saw the greatest revolution in farm mechanisation since man first tilled the land. Most modern-day farm machines have changed beyond recognition during the last seventy years.

Although the reversible plough, reinvented in the late 1940s, was no more than a reincarnation of the horse- and steam-powered balance ploughs at work on farms in the mid-1850s, a late 1940s farming report suggested that the past decade could well have seen the coming of a second agricultural revolution. This prediction proved to be accurate with much of the credit going to Henry Ford, Harry Ferguson, David Brown and other pioneers. Another report recorded the fact that there were thirteen horses for every tractor working on British farms in 1939 - but within ten years horses outnumbered tractors by a ratio of only two to one.

This book traces the development of tractors, tillage equipment, drills, fertiliser distributors, manure spreaders and crop sprayers from the days when Fordson Model N tractors were used to pull trailed ploughs, cup-feed drills and field-heap manure spreaders. The progression from those early farm machines to the latest computerised tractor management systems with satellite navigation operated from the comfort of an air-conditioned cab has happened in little more than a single lifetime. Farming folk, many of whom have only recently achieved the status of senior citizen, will remember sitting on a tractor seat with a corn sack over their knees to keep out the cold, walking behind a cup-feed drill or chopping out sugar beet and turnips with a hand hoe.

The horseman has gone, so have most of the farm men and many of the farms that employed them. Engineers have even developed driver-less tractors but as yet have failed to design machines to go with them which never break down. In spite of all the latest technology there will never be a substitute for the farmer's boot.

In a book of this size there is only room for a random sample of the hundreds of implements and machines used to till the land, drill the seed and care for growing crops. A companion volume dealing with the development of harvest machinery from the sail reaper to six-row sugar beet harvesters is already in preparation.

Brian Bell

Suffolk, 2009

Chapter 1

The Farm Tractor

In 1939 there were at least 700,000 horses and fewer than 55,000 tractors on British farms but by 1945 more than 170,000 tractors were in use. Tractors were mainly used with trailed implements in the 1940s, most had a belt pulley and some had a power take-off (pto) shaft. Tractor drivers often worked up a sweat on cold mornings swinging the starting handle and as soon as the engine had fired up it was switched over to paraffin. Petrol was strictly rationed in those days and tell-tale plumes of white exhaust smoke indicated that the engine was still cold and unburned paraffin was finding its way into the sump.

The arrival of the Ferguson TE 20 tractor in 1946 heralded a revolution in tractor design. It became possible to control ploughing depth with a hydraulic lever instead of a hand lever or a screw handle linked to the depth wheel on a trailed plough. Although David Brown supplied an upholstered seat for two people, other tractors in the late 1940s had a metal pan seat on an unforgiving spring mounting. A starting handle was still part of the tool kit but it was only needed when the battery was flat.

An ex-army great coat and maybe a coomb corn sack over the knees were the driver's only protection from the weather until the early 1950s when a lucky few enjoyed the luxury of a canvas-covered or sheet-metal weather cab. The driver's lot gradually improved with the introduction of less draughty cabs, safety roll bars and safety cabs followed by the armchair comfort of a quiet cab complete with radio, air conditioning and an electronic control system.

1.1 The Fordson Standard tractor which cost less to buy than three farm horses but did the work of eight.

Although successful attempts have been made to produce a driverless tractor, the driver is still indispensable in the twenty-first century and likely to remain so. Radio-controlled tractors were demonstrated in the early 1960s and an electronic guidance system, energised by the tractor battery and using a grid of underground cables, was demonstrated at the 1969 Royal Agricultural Show. An electronic control box and sensor head on the tractor picked up signals from the buried cables to control steering and operate implements.

In more recent times satellite navigation systems have been made which can steer the tractor with incredible accuracy back and forth across a field. The latest electronic control units automatically carry out the necessary sequences to operate many different implements and machines. When ploughing, the electronic control unit will lift the plough out of work at the headland, reverse the plough bodies and lower the plough again ready for the next pass across the field.

Although several makes of tractor were in use on British farms in the early 1930s the vast majority were Fordson Model Fs and Ns made either in America or Ireland. The situation changed when the first blue Model N Fordson Standard tractors were made at Dagenham in 1933 and by 1939 three-quarters of the tractors at work on British farms were Fordsons. The water-bath air cleaner, used on early models, was replaced with an oil-bath cleaner in 1937 when the Fordson was painted bright orange.

Another colour change was made at the start of the Second World War, this time to green, the colour used until 1945 when the blue E27N Fordson Major with orange wheels replaced the Model N. By the end of 1946 Fordson Majors accounted for four-fifths of the total production of four-wheel farm tractors made in Great Britain, but changes were at hand.

An arrangement between Harry Ferguson and David Brown resulted in the production of the Ferguson Brown tractor with the Ferguson draught and depth control hydraulic system from 1936 to 1939. Following the famous handshake agreement the Ford Ferguson 9N was built in America from 1939 until 1947 when Ford produced the similar Ford 8N. The American venture

1.2 The Case tractor was one of the few serious competitors for the Fordson in the 1930s and 1940s. The Model C with a top speed of 4½ mph had three forward gears and one reverse. The power take-off shaft was an optional extra.

ended with a lawsuit whereupon Harry Ferguson entered into an agreement with the Standard Motor Company to build his tractor at Coventry in a redundant wartime aircraft factory. The Ferguson TE 20 went into full production in the summer of 1946.

Early Coventry-built Ferguson tractors, designated TE 20 Con, had the same 24 hp overhead valve Continental engine as the 9N built by the Continental Motor Company at Michigan. When the supply of Continental engines came to an end a modified version of the Standard Vanguard motor-car engine was used and by August 1948 output of the TEA 20 was running at approximately 5,500 per month.

Meanwhile, in order to meet the growing demand for new tractors the Ford Motor Company, which had introduced the E27N Fordson Major at Dagenham in 1945, increased production from 120 a day in 1946 to 230 a day by 1948. The Fordson Major had a 25.8 hp petrol/TVO engine, three forward gears and reverse, a transmission handbrake and, of course, a starting handle. There were four versions: standard, rowcrop, utility and industrial. The standard agricultural tractor on steel wheels cost £237 in 1945 or £285 on pneumatic tyres and the rowcrop model was £255. An optional factory-fitted Perkins P6 diesel engine became available in 1948.

By 1948 agricultural workers' wages were double the rate paid in 1939 and the UK tractor population had reached the 250,000 mark but there were still twice as many horses as tractors on British farms. What's more, the tractor total included almost 13,000 crawlers over 6 hp and approximately 26,000 one- and two-wheel pedestrian-controlled garden tractors used by the nation's smallholders.

Ford and Ferguson did not have it all their own way, as Allis-Chalmers, David Brown, Marshall, Massey-Harris, Minneapolis Moline and Nuffield were all making or selling tractors in Great Britain in 1948. The David Brown VAK 1 with a 25 hp petrol engine and four-speed gearbox had been introduced in 1946; power take-off and hydraulics were optional. Annual output reached 3,500 in 1947 and even more were made in 1948 when David Brown introduced a direct-injection diesel-engined tractor with a high/low range gearbox.

The Series 1 Field Marshall, which replaced the earlier 20 hp Marshall Model M in 1945, had a 40 hp single horizontal cylinder two-stroke diesel engine and diff-lock. The Series 2 followed in 1947 when the diff-lock had disappeared in favour of independent rear-wheel brakes. Marshalls at Gainsborough became associated with John Fowler & Company at Leeds in 1948 and in the following year they

1.3 The Series IIIA Field Marshall was the first Marshall tractor with optional electric starting and hydraulic linkage.

1.4 Caterpillar D2 tractors were imported from America in the late 1940s.

introduced the 40 hp two-stroke engined Fowler Mk VF crawler which cost £1,060.

Other late 1940s tractors included the popular American Allis-Chalmers Model B with a 24 hp petrol/paraffin engine, which was made at Southampton from 1948. The Nuffield Universal M4 with a vaporising oil engine and the petrol-engined PM4 were launched at the 1948 Royal Smithfield Show. The Nuffield Universal was the first British-built tractor with five forward gears and the rear-wheel track setting was adjusted by sliding the wheels along on their axles.

A few hundred Minneapolis Moline tractors with a 46 hp Dorman or 66 hp Meadows diesel engine were imported from America, initially by MM (England) Ltd from 1946 to 1949 and then by Sale Tilney Ltd until the early 1950s. Other American imports in 1948 included Case and John Deere wheeled tractors and Caterpillar D2, D4 and D6 crawlers.

In 1949 International Harvester made the first Farmall M tractors with five forward gears at their new factory in Doncaster. First made in America in 1939, a number of Farmall M tractors were sold in the UK during the war years. Most of them had steel wheels

1.5 Electric lights and starting, hydraulic linkage and power take-off were optional extras for the E27N Fordson Major.

1.6 Weathershield cabs were supplied with a rear roll-up canvas curtain and on hot days the doors could be lifted off in seconds.

and the high fifth gear blanked off, giving the Farmall M a top speed of 6 mph.

Tractor drivers were expected to work in all weathers, just as horsemen had done for centuries. A canvas canopy was an optional extra for the Marshall Model M and the Field Marshall but few other tractor makers were concerned about basic driver comfort. Specialist companies including Sta-Dri, Scottish Aviation, Sun Trac, the Victoria Sheet Metal Company and Weathershields were making weather cabs in the late 1940s. Weathershields' alloy metal cabs with a safety glass screen and a hand-operated wiper blade were made for Ferguson 20, Fordson Major and David Brown Cropmaster tractors and cost £28.

Smallholding was an important part of British agriculture in the 1940s and 1950s when there was an ever-increasing demand for the 5–10 hp ride-on tractors made by numerous companies including Brockhouse Engineering, Garner, Gunsmith, Oak Tree Appliances (OTA), Newman, Singer and Ransomes. The 6 hp Garner four-wheel light tractor had an air-cooled petrol engine, three forward gears and one reverse and a top speed of 10 mph. The range of mid- and rear-mounted implements for the 6 hp and more powerful 7 hp petrol- or TVO-engined Garner included a plough, disc harrows, hoes and seeder units.

1.7 The Garner light tractor had a 6 hp JAP petrol engine.

1.8 High ground clearance and an underslung toolbar made the Newman WD2 an ideal rowcrop tractor.

1.9 The optional hydraulic unit bolted on to the side of the BMB President used oil from the tractor gearbox.

1.10 Sales literature explained that the rear-engined Bean rowcrop tractor was so easy to handle that the driver would not be tired by teatime and could continue working without fatigue.

Gunsmith and Newman tractors introduced in 1948 and OTA, launched at the 1949 Smithfield Show, had a single front wheel and a mid-mounted toolbar. The four-wheel 8 hp BMB President, introduced at the 1950 Royal Show, had a three forward and one reverse gearbox and adjustable track width, making it an ideal tractor for smallholders and market gardeners. The Stockhold President with an air-cooled Petter diesel engine superseded the BMB President in 1956 but very few were made.

The Ransomes 6 hp petrol-engined MG 2 crawler with rubber-jointed tracks cost £135 when it was introduced in 1936. Considerable numbers of the MG 2 were sold to British farmers and fruit growers and the tractor was also popular in French vineyards. An improved MG 5 was introduced in 1948 and the MG 6 with a 7 hp petrol or TVO side-valve engine and a three forward and three reverse gearbox made its debut at the 1953 Smithfield Show. An optional two-stroke diesel engine was available for the MG 40, which superseded the MG 6 in 1960. The last Ransomes MG crawlers were made at Ipswich in 1966.

Self-propelled toolbars were popular in the late 1940s. The three-wheel tiller-steered Bean rowcrop tractor was introduced in 1946. Its 8 hp Ford engine with electric starting and the three forward and one reverse gearbox were mounted above the rear axle. A four-wheel version of the Bean rowcrop tractor was introduced at the 1950 Smithfield Show.

The 14 hp David Brown 2D self-propelled toolbar introduced at the 1955 Smithfield Show was basically a tool carrier with an underslung toolbar, a rear-mounted air-cooled two-stroke diesel engine, a four forward and one reverse gearbox and power take-off. An engine-driven compressor supplied compressed air to the tractor's twin airlift cylinders used to raise and lower the toolbar.

Pedestrian-controlled garden tractors enabled smallholders and market gardeners to get on with their work without the frustrating wait for a local farmer to find time to plough or cultivate their land. The British Anzani Iron Horse, BMB Plow-Mate, Auto-Culto and Trusty two-wheel tractors were cheaper than their four-wheel counterparts and, unlike horses, required no attention when they were not at work. Pedestrian-controlled rotary cultivators

1.11 The David Brown 2D rowcrop tractor was made between 1956 and 1961.

1.12 The Turner Yeoman of England was one of the first British-built diesel-engined tractors.

with single-cylinder engines, including the Howard Gem, Clifford and Simar Rototiller, were also popular with smallholders and farmers growing vegetable crops.

Spark-ignition engines gradually gave way to the four-stroke diesel engine in the early 1950s. The Field Marshall with its two-stroke diesel engine was already popular with threshing contractors but diesel-engined farm tractors were something new. First made at Wolverhampton in 1949 the Turner Yeoman of England diesel tractor, with a 40 hp V4 engine and a four forward and one reverse gearbox, had a top speed of 13 mph.

Launched at the 1951 Royal Show the 25 hp Ferguson TEF 20's diesel engine was started with the aid of a de-compression lever on the dashboard and a Ki-Gas system was provided for use on really cold mornings. Ferguson owners were able to buy a Perkins P3 diesel engine conversion for existing TEA and TED 20 tractors. The grey and gold Ferguson FE 35 replaced the little grey Fergie in 1955 and although most of these tractors left the factory with a 37 hp four-cylinder diesel engine it was still possible to buy an FE 35 with a petrol, TVO or lamp oil engine.

The 50 hp Massey Ferguson 65 with disc brakes and live power take-off made its debut at the 1957 Royal Smithfield Show, with power steering offered as an optional extra. Advertisements at the time explained that by using the 35 and 65 farmers could enjoy 100 per cent Massey Ferguson farming. It was an unlikely coincidence that the Fordson Dexta with a draught control hydraulic system was launched as a 'Workmate for the Fordson Major' in the same year as the Massey Ferguson 65.

Having entered the tractor market with the VAK 1 in 1939, David Brown introduced the Cropmaster in 1947 and a diesel-engined Cropmaster followed in

1.13 The grey and gold Ferguson FE 35 had a four-cylinder diesel engine.

1949. The more powerful Cropmaster 50 with a six-cylinder direct-injection diesel engine and a six forward and two reverse gearbox appeared in 1952. Weighing just over 2½ tons, it was the heaviest British-made wheeled tractor available at the time.

The David Brown 25, 30C, 30D and 50D wheeled models, together with the 30TD and 50TD tracklayers, replaced the Cropmaster and Trackmaster series in 1953. The specification included a twin-range gearbox with six forward and two reverse gears and a dual speed power take-off. David Brown's Traction Control Unit (TCU) used to transfer some of the implement weight on to the tractor appeared in 1954 but implements still needed wheels to control the working depth.

Since the early days of mechanised farming American-built International Harvester tractors have been used on British farms. The first British-built Farmall M tractors were made at Doncaster in 1949, the diesel-engined Farmall BMD was added in 1952 and in the following year they were upgraded to the Super BM and Super BMD. The last Farmall tractors were made at Doncaster in 1958.

1.14 David Brown tractors with their hunting pink paintwork were made at Meltham in Yorkshire.

1.15 The International Harvester B 450 had a 55 hp engine, five forward gears and one reverse and a diff-lock.

The first small British-built International Harvester tractors were made at Bradford in 1955. The 30 hp B 250 had an indirect-injection diesel engine, diff-lock, disc brakes and a live hydraulic system. The more powerful 53 hp B 275 and 55 hp B 450 introduced in 1958 cost £623 and £790 respectively.

Farmers buying a new crawler tractor in the late 1950s had plenty of choice with a dozen or so different makes available, ranging from the 7 hp Ransomes MG 6 to the 65 hp Fowler Challenger II. Bristol Tractors, David Brown, County, Fiat, International Harvester,

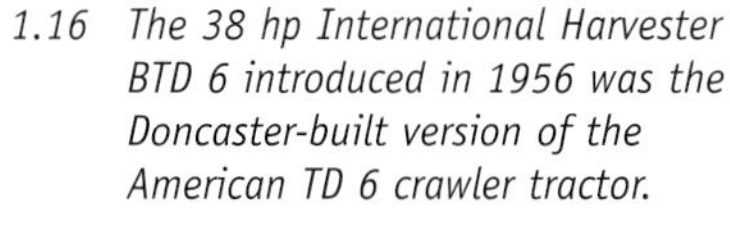
1.16 *The 38 hp International Harvester BTD 6 introduced in 1956 was the Doncaster-built version of the American TD 6 crawler tractor.*

Howard and Roadless were also making crawler tractors in the late 1950s.

County Commercial Cars made a crawler tractor based on the Fordson E27N Major from 1949 with the County Model Z built round the New Fordson Major replacing it in 1952. The County Ploughman, which superseded the Model Z in 1958, remained in production until 1964. County and Roadless Traction also made half-track conversions for several makes of tractor including the Massey-Harris 744D, Fordson E27N Major and New Fordson Major.

Rotaped tracks, used mainly for drainage work, were marketed by Geo Monro in the late 1940s and by Leeford (London) Ltd in the 1950s. They were very different from the usual half-track design and consisted of six linked track sections pulled round a central disc by a system of chains and sprockets.

Four-wheel drive tractors, which appeared in the mid-1950s, were the first real challenge to the virtual monopoly that crawlers had enjoyed for heavy cultivation work since the steam age. County and Roadless four-wheel drive tractors were based on the New Fordson Major. The 52 hp County Four-Drive introduced in 1954 had four equal-sized wheels with clutch and brake steering. The Roadless Manuel four-wheel drive version of the New Fordson Major had small front wheels.

The 100 hp Doe Triple D with its two Fordson Power Major Diesel tractors coupled in tandem was a considerable rival to the tracklayer when production got underway in 1958. A single set of controls operated both engines and the two tractors were

1.17 *The New Fordson Major with Roadless half-tracks. A selling point was said be the ability to change back to pneumatic tyres for rowcrop and haulage work but it was not specified how long it would take to carry out this task.*

1.18 The first Doe Triple D or Doe Dual Drive tractors made in 1960 were based on two Fordson Power Major tractors.

linked by a turntable which allowed the front half to swivel almost at a right angle to give a surprisingly small turning circle.

The 1960 Triple D, based on two Fordson Super Major tractors, had an improved gear-changing arrangement with hydraulic slave rams to enable the driver to change gear without leaving the seat. The Doe 130 appeared shortly after the launch of the Ford 1000 series tractors in 1965. The Doe 150 based on two of the more powerful 75 hp Ford 5000 tractors superseded the 130 in 1968.

Most major tractor companies only made two or

1.19 The David Brown 1200 Selectamatic with a 67 hp direct injection engine, six forward and two reverse gears and independent power take-off was launched in 1967.

three different models in the early 1950s but within ten years most had extended the range. Some were making four-wheel drive models designed on the drawing board rather than by fitting front-wheel drive conversion kits to two-wheel drive tractors.

In 1962 there were eleven different models of David Brown two-wheel drive tractors, ranging from the small 850 to the most expensive 990 with an automatic gearbox. The familiar David Brown red – or hunting pink – gave way to a new chocolate and white livery in 1965. Annual production reached the 30,000 mark in 1967 when the David Brown range included the 770, 880, 990 and the 67 hp 1200. Negative-earth vehicle electric circuits were introduced in the mid-1960s with the David Brown 880 Implematic the first agricultural tractor to have this system.

The launch of the Ford 1000 series tractors coincided with the move from Dagenham to Basildon in 1964. The specification of the new two-wheel drive 2000, 3000, 4000 and 5000 tractors matched the competition with live hydraulics and power take-off, diff-lock and an eight forward and two reverse dual-range gearbox. An automatic gearbox was optional for the David Brown 990 and Massey Ferguson tractors could be supplied with a multi-power transmission.

However, Ford topped the lot with their optional Select-o-Speed change-on-the-move transmission. It had ten forward and two reverse gears engaged with a single lever and an inching pedal was provided for implement hitching and controlling the tractor in very confined spaces.

John Deere re-entered the British wheeled tractor market in 1963 when forage harvester manufacturer Lundell imported the 84 hp 4010 and 125 hp 5010. John Deere, which acquired Lundell in 1965, established a depot at Langar near Nottingham in 1967. There were five John Deere tractors on the British market in the late 1960s, ranging from the 47 hp 1020 to the 143 hp 5020 with lower link sensing and closed centre hydraulics.

Massey Ferguson introduced the Perkins-engined 130, 135, 165 and 175 tractors in 1965. Optional equipment, although not available for the 130, included a purpose-built cab, multi-power, hydraulic

1.20 The Ford 4000 was one of the new 6X or 1000 series tractors made at Basildon from 1964.

1.21 *This John Deere 4020 tractor is hitched to a Stanmill chisel plough.*

spool valves and a foot throttle. Pressure control, a system to transfer some of the weight of a trailed implement on to the rear wheels of the tractor, was also an optional extra.

The Nuffield Universal M tractor announced in 1948 was joined by the smaller Nuffield Universal 3 in 1957. Both tractors were made at Birmingham until the late 1950s when production moved to Cowley. The four-cylinder 60 hp Nuffield 4/60 was an improved version of the earlier Universal 4. Production moved to Bathgate in Scotland in 1963 and shortly after this the 4/60, now with a ten forward speed gearbox, became the 10/60, with the improved Nuffield 3 redesignated the 10/42.

The BMC Mini Tractor, introduced at the 1965 Smithfield Show, had a 15 hp diesel engine, a nine forward and three reverse speed gearbox and optional hydraulic system. The Nuffield 3/45 and

1.22 The Massey Ferguson 165 was one of the Red Giant 100 series introduced in 1964.

1.23 The Nuffield 4/65 had a 65 hp four-cylinder engine.

4/65 replaced the 10/42 and 10/60 in 1967 and in the following year BMC became part of British Leyland. The Nuffield name and orange paint disappeared in 1970 when the tractors were re-styled, painted blue and given a Leyland badge.

British Leyland then acquired Marshall-Fowler but the enlarged company went through troubled times and in 1979 was bought by Charles Nickerson who gave the wheeled tractors a facelift and launched the new gold and black Leyland models in 1980. Nickerson also revived the Marshall and Track Marshall companies and both were exhibited at the 1982 Smithfield Show when the Marshall TM120, TM135 and the 70 hp Britannia crawler were the only British-made tracklayers at the event. Eight models of Marshall wheeled tractor, previously sold with Leyland badges, were also shown.

1.24 There were two- and four-wheel drive versions of the 103 hp black and gold Marshall 100 tractor.

1.25 The Muir Hill 101 had a six-cylinder Ford diesel engine.

The decline of crawler tractors from the mid-1960s was mainly due to the growing popularity of four-wheel drive tractors, including the County 1124, Doe Triple D, Muir Hill 101 and Roadless Ploughmaster. Most of them had a Ford engine and many used Ford transmission components.

Tractors became more powerful and more expensive in the 1970s when the larger four-wheel drive models included the 163 hp Muir Hill 161 and the Roadless 115. Seven years later the 325 hp Ford FW30, really a Steiger in Ford Blue livery, was by far the largest and, at £49,109, the most expensive wheeled tractor on the British market.

Famous tractor makers including David Brown, Ford, International, Leyland and Nuffield lost their independence or completely disappeared during the 1970s and others such as Belarus, Deutz, Fiat, Renault, Valmet and Zetor took their place. The first

1.26 Introduced in 1972, the articulated Massey Ferguson 1200 Multi-power transmission had twelve forward and four reverse gears.

1.27 *The 335 hp four-wheel drive Ford FW 60, made by Steiger in America, was the largest and most expensive tractor on the British market in 1985.*

1.28 *The Marshall TM rubber-tracked crawler had a 210 hp turbocharged Cummins diesel engine.*

turbo-charged Ford tractor, the 7000, was launched in 1971 and by the late 1970s Ford TW four-wheel drive tractors were on the market.

Tenneco, which already owned Case, acquired David Brown in 1972 and changed the model numbers of the tractors; for example, the David Brown 1470 became the David Brown Case 2470 Traction King. The International Harvester tractor division also became part of the Tenneco empire in 1985 and a new generation of Case IH Magnum tractors was introduced to the UK in 1989. Earlier in 1986 Case IH bought Steiger which built high horsepower articulated four-wheel drive tractors in America.

Ford New Holland were selling Steiger tractors badged as the Ford FW series in the UK and the change of ownership left Ford New Holland without a big four-wheel drive tractor. The acquisition of the Versatile Farm Equipment Co in 1987 overcame the problem but sales of 300 hp-plus tractors were in decline and only a few Ford Versatile tractors were made on special order.

1.29 The JCB Fastrac with an eighteen forward and six reverse speed gearbox was launched at the 1990 Smithfield Show.

Crawler tractors were disadvantaged by their low top speed and unsuitability for roadwork until the arrival of the 270 hp rubber-tracked Caterpillar Challenger 65 in 1989 and the 210 hp Track Marshall TM 200 which made its debut in 1990.

The JCB Fastrac high-speed tractor appeared in its pre-production form at the 1990 Smithfield Show. The first production models with coil-spring front suspension system, self-levelling hydro-pneumatic rear suspension and air-operated disc brakes went on sale in the summer of 1991. Early Fastracs had a 125 hp naturally aspirated or 145 hp turbocharged Perkins diesel engine and an eighteen forward and four reverse gearbox.

Improved wheel equipment and gearing introduced at the 1992 Smithfield Show gave the Fastrac a top speed of 40 mph and the hydraulic linkage was redesigned for category III implements.

Chapter 2

Ploughs

The mouldboard plough has been the primary farm tillage implement for many centuries but in more recent times minimal cultivation techniques and direct drilling have challenged its supremacy. However, persistent perennial weeds and drainage problems in some soils have ensured its survival to the present day.

Trailed ploughs have been used since tractors first appeared on the farm and they were still being used with crawler tractors long after the Ferguson tractor hydraulic system arrived on the scene. The first trailed ploughs were chain-pulled by a steam engine or an internal combustion-engined tractor. The tractor ran on

2.1 Two horses with a single-furrow Ransomes YL ploughed an acre a day.

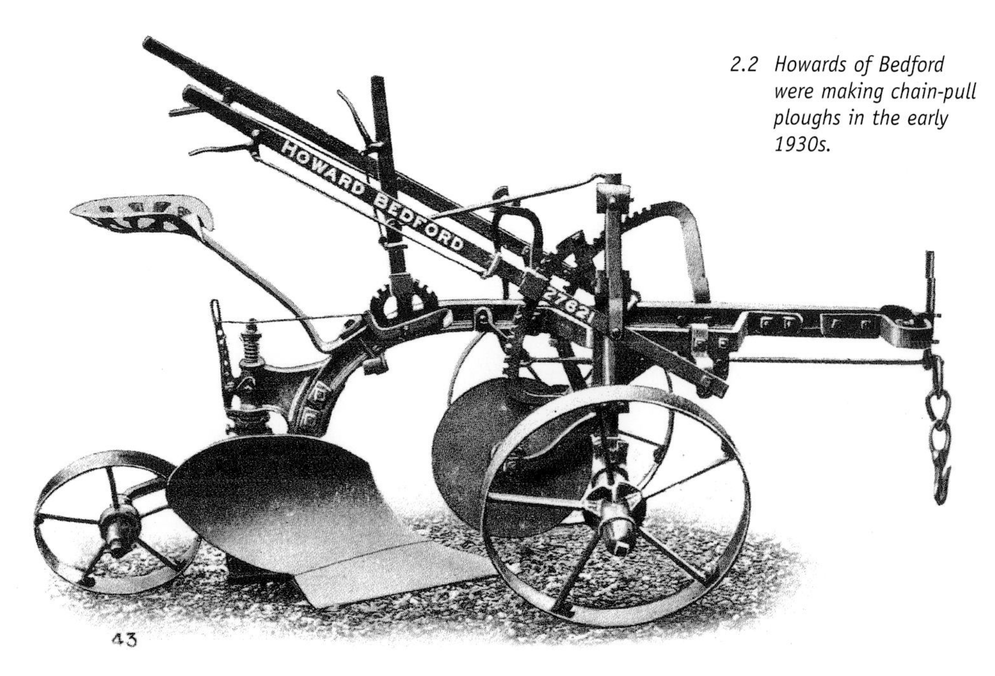

2.2 *Howards of Bedford were making chain-pull ploughs in the early 1930s.*

the land and a seat was provided from which a man could steer the plough and lower it into or lift it out of work.

The first Ransomes chain-pull plough with a deep digging body appeared in 1904, a steerable two-furrow model was added in 1909 and the three- and four-furrow steerable RYLT (Ransomes Yorkshire Light Tractor) appeared in 1914.

One-man tractor ploughing dates back to 1919 when Howards of Bedford, Ransomes and others made trailed ploughs with a rope-operated rack-and-pinion self-lift mechanism. The bodies were taken out of work by pulling a cord to engage a curved-toothed rack with a sprocket on the land wheel hub to lift the plough. Another tug on the cord dropped the plough back into work.

2.3 The two-furrow Weetrac, introduced in 1927, was the first Ransomes mounted plough.

Like most trailed ploughs of the day, the Ransomes RSLD (Ransomes Self-Lift Double) or RSLM (Ransomes Self-Lift Multiple) were used with the right-hand tractor wheels running in the furrow. However, some farmers preferred to run the tractor wheels on unploughed land and to meet this demand Ransomes also made the centre-lift RCLD and RCLM steerable ploughs.

The introduction in 1946 of the Ferguson TE 20 tractor with its three-point linkage mounted plough was the beginning of the end for the trailed plough. The origin of the mounted plough dates back to 1917 when a two-furrow tractor-mounted plough without any wheels was made in America by Harry Ferguson for the Ford Eros tractor.

A simple mechanical linkage was used to attach the plough to the tractor and a hand lever was provided to lower and lift the plough into and out of work. To help overcome the tendency for the front of the tractor to rear up when under a heavy load the plough hitch point was positioned under the tractor and forward of the rear wheels.

In the mid-1920s a modified version of the Eros plough was made in America for the Fordson tractor. The similar Ransomes Weetrac mounted plough with two-furrow general-purpose bodies was introduced in 1927 for the mechanical lift linkage on Fordson and other tractors.

A Ransomes implement catalogue for 1932 illustrated fifteen models of trailed tractor plough and twice as many one- and two-furrow horse ploughs, including balance and turnwrest reversible ploughs. Fewer horse ploughs were included in the Ransomes catalogue for 1939 when tractor ploughs included the Motrac, RSLD and RSLM, Duotrac, the five-furrow Quintrac and the 31 cwt six-furrow Hexatrac.

A self-lift mechanism, operated through the land wheel and a cranked axle, was an essential part of trailed ploughs. Some plough makers used a hub and roller lift clutch, others preferred the rack-and-pinion mechanism also used on cultivators and other trailed soil-engaging implements. The trip rope was attached to the tractor seat with a quick-release spring clip to prevent it being pulled from its mounting if the plough came to a sudden halt. Most ploughs had a wooden shear peg or spring-loaded safety hitch built into the drawbar to protect the share from damage if it hit a tree root or other underground obstruction.

Several companies, including Cockshutt, Fisher Humphries, the Ford Motor Co, International Harvester, Massey-Harris, Ransomes, Ruston & Hornsby, Oliver, Sellar, Talbot and Wilmot, were making trailed ploughs in the 1940s. Home-grown food supplies were vital during the war years and the manufacture of farm machinery was just as important as making armaments for the war effort. Plough

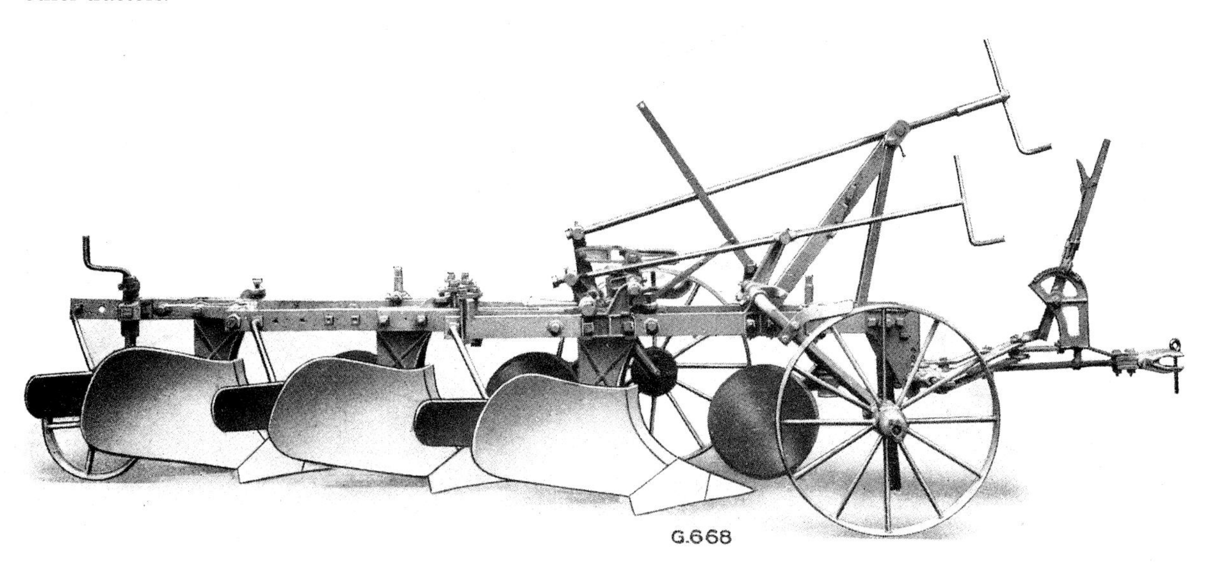

2.4 Made by Ruston & Hornsby at Grantham in the late 1930s, this semi-digger Hornsby plough has a rack-and-pinion lift mechanism.

2.5 The two- and three-furrow Fordson Elite ploughs were made at Leamington Spa.

production was a top priority and in 1943 Ransomes made Motrac No 3 ploughs at a rate of 130 a week.

With hostilities at an end the Ford Motor Co made the first two- and three-furrow Fordson Elite trailed ploughs with Ransomes bodies at Leamington Spa in 1945. Three years later the first International Harvester B-8-A tractor ploughs were made at Doncaster.

The two- and three-furrow McCormick International B-8-A ploughs for the Farmall M tractor could be set to plough 9–12 in wide furrows and were typical of the type. Two hand levers, connected to the wheels, were used to set the depth and level the plough. Front furrow width was adjusted with a lever on the drawbar and a spring-loaded safety hitch released the plough if it hit an obstruction. The Massey-Harris No 28 five-furrow semi-digger plough had an oil-bath lift clutch which, when the plough was out of work, was claimed to maintain the bodies at a height of 7½ in above the ground automatically. The single-furrow Deepacre and Bracre and the 4–3 furrow Moracre were three of the trailed ploughs made by Fisher Humphries at Pershore in the 1940s and 1950s.

2.6 Massey-Harris advertised the new M-H No 28 five-furrow plough in 1945.

2.7 *SM Wilmot & Co made the single-furrow Turnall plough at Bristol.*

Introduced in 1948, the Deepacre, advertised as 'the plough for tomorrow for the farmer of today', was a single-furrow deep digger plough and, depending on the type of body used, turned a furrow either 14 or 18 in deep. A full range of plough bodies, including a long and narrow lea type for grassland and deep digging bodies, was made for the Moracre plough. The Bracre was used for land reclamation and breaking up virgin soil.

The Wilmot Turnall, advertised in the early 1940s as a general-purpose or deep digger plough, had a hinged draw plate bolted to the tractor drawbar. The leg and body were hinged at the back of the beam and a small rear wheel was provided for transport. A hand lever at the front of the plough lowered the plough into work and, on reaching the required depth pre-set with a screw handle, the plough levelled itself out and remained at that depth.

The beams on the later Wilmot 'Quick-Adjustable' mounted two-furrow plough were held parallel and adjusted for furrow width with swivel cross members. The hydraulic draught control lever and the position of the hitch on the plough frame controlled ploughing depth.

2.8 *The Wilmot 'Quick-Adjust' plough was first seen at the 1949 Royal Show.*

2.9 The lift clutch on the Fisher Humphries Moracre plough was linked to the rear furrow wheel.

General-purpose, semi-digger and digger bodies were available for most ploughs but the famous Ransomes YL (Yorkshire Light) general-purpose body reigned supreme for many decades. However, the appearance of continental ploughs in the 1950s with their new-style mouldboards suitable for both high-speed shallow ploughing and deep work heralded the end of the traditional general-purpose body.

Reversible or one-way ploughing with a balance plough dates back to the steam age. Horse-drawn reversible ploughs were in use in the early 1900s but tractor-drawn reversible ploughs were not popular until the introduction of the Ferguson 'butterfly' reversible plough in the late 1940s. The trailed Hosier-Bomford reversible had three pairs of bodies attached to a box-section beam mounted on a two-wheeled carriage at the front and a second pair of wheels at the rear. The wheels on one side of the plough ran in the furrow while the others were on unploughed land.

Fordson tractors with trailed Ransomes ploughs did most of the ploughing on arable farms during the immediate post-war years. However, the arrival in 1946 of the Ferguson TE 20 with its hydraulic depth control system and ploughs to match was a turning point in the history of farm mechanisation. The Ford Motor Co, International Harvester, David Brown and others were busy developing their own systems but the Ferguson patents meant that they were restricted to simple lift-and-drop hydraulic systems with a depth wheel an essential part of the plough.

'Don't say plough, say Ransomes' was a slogan of this famous old Ipswich company which made many

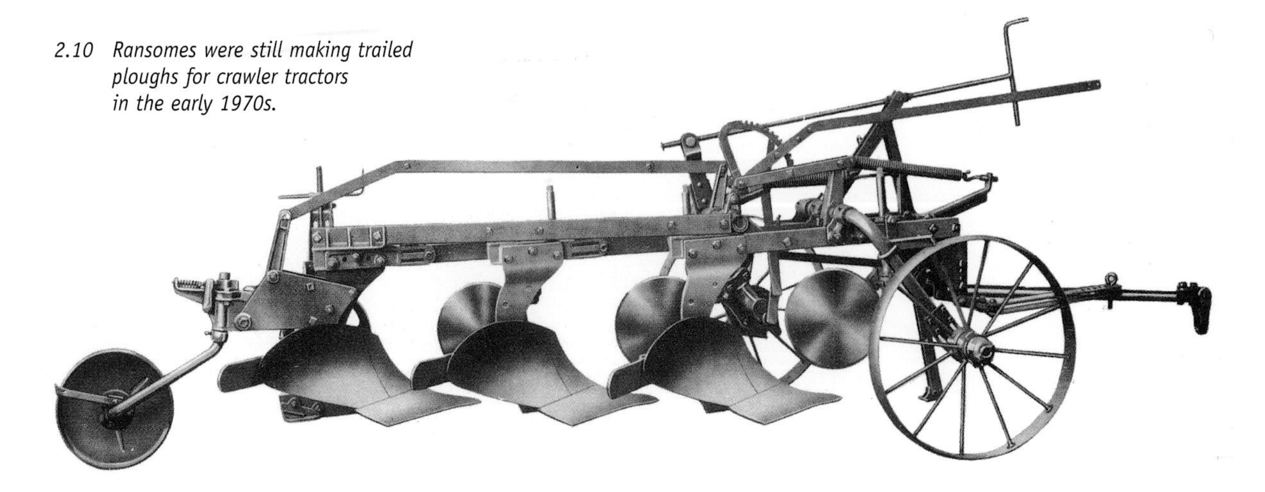

2.10 Ransomes were still making trailed ploughs for crawler tractors in the early 1970s.

2.11 The Ransomes FR two-furrow TS63 with a depth wheel was made for tractors without a draught-control hydraulic system.

different types of plough in the mid-1950s and enjoyed the lion's share of the market both at home and abroad. The Ransomes catalogue for 1956 included fifteen different mounted, trailed and reversible tractor ploughs with the choice of ten body types. Other plough makers at the time included Bonnel, David Brown Albion, Fisher Humphries (which was acquired by AB Blanch at Crudwell in 1956), McCormick International, Kverneland, Massey Ferguson, Melotte, Pierce and Geo Sellar.

Reversible or one-way ploughing became popular in the early 1950s when Barford, Bonnel, Ferguson and Ransomes, to name but a few, were making progress in this growing market. This was in spite of comments often heard from wise old tractor men who saw little point in having an expensive set of plough bodies stuck up in the air doing nothing!

Continental Farm Equipment at Salhouse in Norfolk imported a single-furrow semi-mounted Bonnel reversible plough from France in 1947. Within a year or so single- and two-furrow semi-mounted Bonnel ploughs with soil-wearing parts

2.12 Ferguson mounted ploughs had a rolling rear landside with a scraper to keep it clean.

2.13 The Bonnel two-furrow reversible plough was approved for use with McCormick International wheeled tractors.

2.14 The Ferguson single-furrow plough was reversed by a simple trip mechanism when it was lifted out of work.

2.15 The Norfolk Bonnel reversible plough, designed for crawler tractors, was lifted automatically when the tractor turned at the headland.

2.16 The Ransomes TS 51 reversible plough had a manual turnover mechanism.

2.17 The single-furrow David Brown reversible plough for the 2D and VAK tractors consisted of separate left- and right-hand ploughs mounted side by side.

2.18 The Webb 'Brenig' reversible plough bodies rolled over in the direction of travel.

from France were made at Salhouse. The plough was rigidly hitched to the tractor drawbar and carried on two castor wheels at the rear. A certain amount of muscle power was required when using a handle to reverse the bodies ready for the next run.

The Bonnel mouldboard was popular with East Anglian farmers and in 1953 Continental Farm Equipment introduced the three-furrow Norfolk Bonnel with an automatic turnover mechanism for crawler tractors.

Most reversible ploughs had a manual turnover mechanism at the time, some being quite hard work to use. The Ferguson single-furrow 'butterfly' reversible introduced in 1951, had an automatic chain-operated turnover mechanism attached to the back of the tractor. Improved manual turnover mechanisms on Lemken and other continental reversible ploughs, which used the weight of the plough, made it easier to reverse the bodies at the headland.

The David Brown single-furrow reversible plough was really two separate left-hand and right-hand ploughs, attached with a special hitch arrangement to the three-point linkage. A wire rope running over a pulley on the back of the tractor, which linked the left- and right-hand plough beams, was used to raise one body from work and lower the opposite one as the tractor re-entered the furrow for the next run.

Farmers were able to compare the work done by Bamford, Bomford, Bonnel, Doe, Lemken, International Harvester, Massey Ferguson, Melotte, Ransomes, Salopian Huard and Webb reversible ploughs at a Cambridgeshire Farm Machinery Club demonstration in 1962. Most were mechanically reversed, three had a hydraulic ram and three more needed the assistance of the tractor driver's arm.

The two-furrow Webb Brenig plough was unusual with the bodies reversed by rotating them in the direction of travel. The hydraulically reversed Doe four-furrow reversible plough with Ransomes bodies, demonstrated on a Doe Triple D, cost £525, more than twice the price of any other plough at the demonstration.

This new generation of reversible ploughs, which followed the earlier one-way horse ploughs and steam-powered balance ploughs, eventually saw most farmers consign their right-handed ploughs to the scrap merchant or nettle bed. However, there was a small hiccup in the early 1970s when the available tractor power overtook the design of reversible ploughs. This left the way clear for a brief comeback on large arable farms of several five-, six- and seven-furrow semi-mounted conventional ploughs, which could use the extra power more efficiently.

David Brown, International Harvester, John Deere, Ransomes and others made mounted, semi-mounted and trailed multi-furrow conventional ploughs at the time. Lundell was importing five-furrow semi-mounted John Deere conventional ploughs, which could work a maximum depth of 12 in at speeds of up to 6 mph. The rear land wheel on the five-furrow John Deere and six-furrow semi-mounted David Brown plough was used to control the ploughing depth at the back end of the plough. The fully mounted McCormick International B1-51 five-furrow model was easily reduced down to a four- or three-furrow plough in hard going.

At this time Ransomes still made the

2.19 Early Lemken reversible ploughs had a mechanical turnover mechanism.

2.20 Ransomes introduced the two- to five-furrow TS 90 New Theme ploughs in 1967.

2.21 A Doe tool carrier and four-furrow Doe reversible plough on a Marshall 70 crawler at a 1967 ploughing demonstration.

five- and six-furrow Hexatrac for crawlers and high-powered wheeled tractors. Ransomes mounted ploughs in the mid-1960s included the five-furrow TS 59 and the five-furrow semi-mounted TS 78. The TS 90 New Theme range of two- to five-furrow mounted ploughs with hollow box-section beams and TCN high-speed bodies made its debut in 1967.

The TS 92 with six, seven or eight bodies was added in 1968 and farmers could add extra bodies to these ploughs by obtaining a longer beam under the Ransomes beam exchange scheme.

As there was no effective hydraulic three-point linkage for crawler tractors, ploughing was done with a multi-furrow trailed plough. The situation changed when wheeled tool carriers made by Bomford, Robert Crawford, Ernest Doe & Son and Turner Engineering appeared in the mid-1960s. They were equipped with external rams, making it possible to use standard mounted implements, including hydraulically reversed ploughs, behind a crawler tractor.

Warwickshire farmer Roger Dowdeswell, well aware of the disadvantage of ploughing with crawler tractors without hydraulic linkage, designed a three-point linkage unit for the farm's crawler tractor. After selling the manufacturing rights to Turner Engineering at Alcester he set about building a prototype fully mounted offset reversible plough for the farm crawler.

The four-furrow DP1 (Dowdeswell Plough 1) with Ransomes bodies was attached to the crawler by a swinging linkage which kept the plough in line and was unaffected by the slewing steering action of the tractor. The DP1 plough, which was awarded an RASE Silver Medal at the 1970 Royal Agricultural Show, was still in production in the mid-1990s.

Following his success with the fully mounted reversible plough for crawler tractors Roger Dowdeswell introduced the two- to six-furrow mounted DP5 conventional plough in 1972. In response to a demand for larger ploughs Dowdeswell also introduced the multi-furrow semi-mounted DP2 for wheeled tractors in 1972, while the six- to eight-furrow semi-mounted DP6, also for wheeled tractors was added in 1973.

Launched in 1974, the four- to seven-furrow fully mounted DP7 reversible for wheeled tractors up to 140 hp was the first Dowdeswell plough aimed at the volume market. Plough production soon outgrew the Dowdeswell farm workshop and the business was moved to the Blue Lias Works at Stockton near Rugby in 1975. The lightweight DP8 mounted reversible was launched in 1981 and the six- to eight-furrow semi-mounted DP9 was the first Dowdeswell plough with variable furrow width adjustment.

2.22 *The Dowdeswell DP1 was the first fully mounted reversible plough for crawler tractors.*

Minimal cultivations and direct drilling were becoming fashionable in arable farming areas in the early 1970s. Many farmers used a mouldboard plough, others chisel ploughed the land and a few sprayed off the weeds before sowing their seed corn with a direct drill. Many an argument raged over the need to plough at all.

A number of different makes of heavy-duty cultivator, more generally known as chisel ploughs and used by farmers who considered them to be an alternative to the mouldboard plough, were on the market between the late 1960s and the early 1980s. Chisel ploughs were either made or imported by a dozen or so companies including Alpha

2.23 *The DP7 fully mounted Dowdeswell reversible plough, first made in 1974, was suitable for four-wheel drive and big two-wheel drive tractors.*

2.24 This 1989 rubber-tracked 270 hp Caterpillar Challenger 65 and ten-furrow Dowdeswell semi-mounted reversible plough had an output of 4 acres per hour when working at 4 to 5 mph.

Accord, Bamford, Brown, International Harvester, Kverneland, Lemken, Massey Ferguson, Ransomes, Stanmill, and Vicon.

Lemken chisel ploughs were really heavy-duty cultivators but the Stanmill, with its tines set at a steep angle, was a good load even for the most powerful tractors of the day with around 100 hp under the bonnet.

Chisel ploughs had between five and sixteen rigid or spring tines in two or three widely spaced rows staggered across a heavy-duty frame. The working width, which depended on the available tractor power, was between 6 ft 6 in and 12 ft.

Some farmers who were not in favour of chisel ploughing and considered it unnecessary to deep-plough land for cereal crops used the alternative option of shallow ploughing at high speed with a stubble plough. Shallow ploughing, or riffling as it was known in some areas, was not a new idea; Ransomes and other plough makers were selling stubble ploughs in the late 1920s and early 1930s.

Fuel economy and high work rates could be achieved with a shallow or stubble plough with up to ten bodies working at a depth of about 6 in. This technique was popular on the continent in the late 1970s when Krone, Lemken, Melotte and Rumpstad among others were making stubble ploughs.

In the early 1980s plough manufacturers were building bigger and stronger ploughs in order to match the ever-increasing power of the 250–350 hp

2.25 Massey Ferguson chisel ploughs were made by Bomford & Evershed.

2.26 The Stanmill Scimitar chisel plough was a good load for a 100 hp tractor.

2.27 *Some arable farmers used Lemken and other makes of stubble plough during the late 1970s and early 1980s.*

2.28 *Naud and other makes of push-pull plough failed to make a real impact on British farms in the early 1980s.*

2.29 *A hydraulically operated parallel linkage maintains all of the furrows at the same width on this variable furrow width Kverneland plough.*

2.30 *Automatic steering enabled the Dowdeswell DP9 semi-mounted plough to follow the tractor wheel tracks.*

2.31 *The first articulated reversible ploughs imported from France in 1990 had a self-steering central axle. The front and rear sections were independently lowered and lifted into and out of work.*

four-wheel drive tractors from North America. However, the giant tractor boom soon passed and most farmers settled for a tractor in the 100–150 hp bracket and used a five-, six- or seven-furrow reversible plough. Plenty of front-end weight was needed on the tractor to counterbalance these heavy ploughs.

Cast-iron front weights were expensive and, with soil compaction causing concern, push-pull ploughs with mounted reversible ploughs on the front and rear hydraulic linkages appeared in 1982. The idea of carrying useful front weight rather than heavy cast-iron weights attracted considerable interest which peaked in 1985 when Fiskars, Lemken, Rabewerk

2.32 *The Howard reversible square plough was reversed with a ram which moved the swinging beam through an arc to bring the opposite end of the mouldboards into work.*

2.33 Offset headland ploughs, including those made by Twose at Tiverton in the 1950s, were popular with farmers who wanted to plough close up to the hedge to control weeds and gain some extra furrows around the field.

and Ransomes push-pull ploughs attracted much attention at that year's farm machinery demonstrations.

Although they were still made in the early 1990s their popularity was short-lived This was partly due to the difficulty in matching the work of the front and rear ploughs but also because of the problems encountered when negotiating narrow country roads.

Variable furrow width ploughs, which appeared in the late 1970s, made it easier to achieve the highest possible output in all field and weather conditions. All the furrow widths, with the exception of the front furrow, were changed with a single adjustment but it was still necessary to re-set the front body to match. The problem was solved in 1983 when Kverneland introduced the Vari-Width system. A hydraulic ram operated a special linkage system which was used to alter the width of all the furrows even while ploughing was in progress.

Push-pull ploughs failed to make any real impact on British farming and by the late 1980s farmers wanted bigger multi-furrow reversible ploughs to match their 100 hp-plus four-wheel drive tractors. Most plough makers met this need by adding more furrows to existing mounted models but wider headlands were needed to turn with such long ploughs.

Articulated reversible ploughs, first imported from France in 1990, were expensive but, provided that the farmer had a sympathetic bank manager, they solved the manoeuvrability problem. With the front of the plough on the tractor linkage and the rear section articulating about a central steering axle, combined with independent lifting and lowering of both sections, headlands could be narrower than they were for a push-pull plough.

Over the years, plough manufacturers spent vast sums of money in the search for the perfect mouldboard. Then, in 1989, the reversible square plough arrived from America where it was used to break up hard, dry stubble. With double-ended mouldboards, similar to short bulldozer blades, the square plough made a big initial impression in arable farming circles. The bodies were reversed by swinging the beam through an arc with a hydraulic ram and many square ploughs had mouldboards with durable plastic facings to extend their working life.

Square ploughs worked well for the first three or four years while Britain enjoyed hot dry summers but the next harvest was very wet and, as square ploughs did not work in wet conditions, it was back to the mouldboard.

Square ploughs and chisel ploughs are not the only

2.34 At an average digging speed of 1 mph the 7 ft wide Vicon Rotaspa spading machine had a work rate of up to three quarters of an acre in an hour.

implements that have appeared over the years in an attempt to make the mouldboard plough obsolete. Other ideas included the 1957 German Schraubpflug (or screw plough) with a pto-driven auger which 'ploughed' a well-broken furrow 10 in wide and 8 in deep.

The Raussendorf Kombinus rotary plough, imported by Manns of Saxham in 1962, had two short mouldboards, each with an offset vertical pto-driven cutting rotor used to turn and pulverise the soil.

The green-and-yellow Rotagri spading machine, made in Belgium and sold by Thompson & Stammers at Dunmow in Essex in the mid-1950s, mimicked the action of hand digging.

The spading machine became the Rotaspa in 1959 when Vicon made it in Holland. The three-point linkage mounted and pto-driven Rotaspa, with a working speed of 1 mph, had eighteen spades in groups of three spaced across its 7 ft digging width. A similar digging machine made by Fahr in Germany in the mid-1950s was said to plough, cultivate and harrow up to 8 in deep in a single pass.

Chapter 3

Cultivators and Harrows

Fashions in cultivation implements have changed many times since the 1930s. After a slow start, while the tractor gradually replaced the horse, the revolution in tillage implements was finally triggered when power take-off and three-point linkage became standard equipment on farm tractors. It was progress indeed when a new range of tillage implements for the Ferguson automatic depth-control hydraulic system appeared in the mid-1940s.

Mounted implements were also made for other makes of farm tractor but their basic lift-and-drop hydraulic systems required one or more wheels to control the working depth. Later generations of tillage machinery included various types of rotary cultivator, power harrow and tillage train.

Cultivators

Over the years, farmers have used horse-drawn, tractor-trailed or mounted, rigid or spring-tined cultivators to break down ploughed land as the first stage of seedbed preparation. Although mounted cultivators were in widespread use by the early 1950s a number of companies, such as Bettinson, Leverton, Martin and Ransomes, continued to make trailed models.

WE Martin, who won an RASE prize for a horse-drawn cultivator in 1900, founded the Martin Cultivator Co at Stamford in 1904. Within ten years Martin products included hay machinery, potato equipment and trailed cultivators. The seven-, nine- and eleven-tined rigid and spring-tined tractor cultivators made by the company in the late 1940s had a lift mechanism on the land wheels and a screw handle to adjust cultivating depth. Ridging bodies were made for Martin cultivators and the width between the wheels on the Martin Macrop cultivator could be adjusted for working in different width rows of crops.

Markham Traction Ltd, which made farm trailers, bought the Martin Cultivator Co in 1950. Production of Martin-Markham tillage implements, trailers, rotary cultivators and other machinery continued until the late 1960s when the firm changed over to the manufacture of road rollers. Simplex, which acquired Martin-Markham in 1970, closed the business.

Ransomes made horse-drawn cultivators in the late 1800s. Tractor-drawn cultivators were added in the early 1900s and by the mid-1930s they were making self-lift Dauntless and Equitine cultivators and hand-lift tractor-mounted toolbars with cultivator tines.

3.1 Martin tractor cultivators were made at Stamford in Lincolnshire.

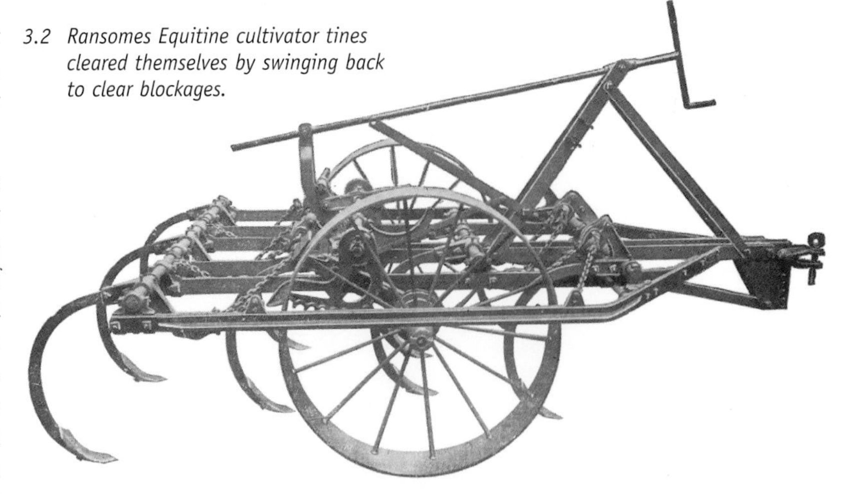

3.2 *Ransomes Equitine cultivator tines cleared themselves by swinging back to clear blockages.*

Standard and heavy-duty Dauntless cultivators had seven, nine, eleven and thirteen rigid or spring-mounted tines.

Ransomes Equitine cultivators had four rows of individually mounted swinging tines connected together by a system of chains. When one of a linked set of tines met an obstruction or was overloaded with trash, the chain linkage allowed it to swing back to clear the blockage. The blocked tine pulled the other tines in the group forward and after the blockage had cleared, the chain linkage returned the tines to their working position. Later models of the Equitine cultivator had groups of tines linked by chains to compensating arms on the tine bars.

The first mounted cultivators, attached to brackets on the tractor rear axle, had wheels to control working depth and a hand lever to put the tines into and out of work. Depth wheels were necessary on three-point linkage mounted cultivators for all makes of tractor apart from rigid, spring-loaded and spring-tine mounted cultivators for the Ferguson draught control hydraulic linkage.

3.3 *This David Brown mounted cultivator or tiller has spring-loaded tines.*

Mounted flexible spring-tine cultivators, made by a number of different companies, were popular with UK farmers in the 1950s and 1960s. Kongskilde, Triple K and Flemstoffe spring-tine cultivators from Scandinavia were first on the scene followed by several British makes including Bentall, Catchpole, Massey Ferguson, Ransomes, Russells of Kirbymoorside and Twose.

The first spring-tine cultivators were typically about 8 ft in width but to keep pace with the increasing tractor power they became wider over the years. Some spring-tine cultivators, with a working width of 20 ft or more, had a hand-operated winch arrangement or hydraulic rams to fold the wing sections for transport while others were end-towed from farm to field on pneumatic-tyred transport wheels.

Various types of seedbed cultivator, with a combination of crumbler rolls, harrow tines and levelling boards added to a spring-tine cultivator, appeared in the early 1970s. These combination cultivators, some with working widths of 25 ft or more for 100 hp tractors, are still in use today.

3.4 Working widths of the Vicon Vi-Till spring- tine cultivator ranged from 6 ft to 11 ft 6 in. Wider models with hinged wings went up to 16 ft 3 in.

3.5 The widest 23 ft 6 in Massey Ferguson MF 37 flexible tine cultivator folded down to 9 ft 10 in for transport.

Harrows

The disc harrow, originally from America, remains an important primary tillage implement. Apart from increased working widths as the horse gradually gave way to the internal combustion engine, disc harrows have remained unchanged for many years. Bamfords, Bentall, Dening, International Harvester, Pettit and Ransomes among others made trailed disc harrows in the 1940s and 1950s.

3.6 The 7 ft 6 in wide Bentall 'Digga' tandem disc harrow which cost £133 in 1960 had hardwood bearings. Four cast-iron transport wheels cost an extra £10.

Detachable cast-iron wheels were provided for transporting disc harrows on public roads but this was necessarily a slow procedure. The introduction of mounted disc harrows solved the problem and in more recent times pneumatic-tyred transport wheels made it much easier to move wider heavy-duty trailed disc harrows from farm to field.

Single-tandem harrows have two gangs of discs with one in front running at an angle to the rear gang. Double-tandem disc harrows have four gangs of discs, which can be angled to vary their effect on the soil. Maximum soil disturbance is achieved when the discs run at an angle and when set to run almost straight ahead they cut and consolidate the soil at the same time.

Each gang, with discs of between 18 and 24 in diameter and spaced 6–9 in apart with cast-iron bobbins, was carried on a square shaft attached to the harrow frame by a pair of cast-iron bearing blocks. Some had wooden bearing blocks, usually of maple, while more expensive harrows were sometimes fitted with roller bearings.

Upon the arrival of the power harrow in the late 1960s many farmers left their disc harrows in the shed. This was not the case in grassland areas where tined implements pulled out the freshly ploughed-in turf, making them unsuitable for seedbed work.

Disc harrows enjoyed a revival during the era of minimal cultivations in the mid-1970s. The new generation of disc harrows with large-diameter discs threatened to replace the plough, but it survived and still shares a role with heavy cultivators, chisel ploughs and disc harrows as one of the primary cultivation tools on British farms.

3.7 The Dening Somerset double-tandem mounted disc harrow with cast-iron bearings could have 18 or 24 in diameter tempered steel discs.

3.8 Ransomes HR 46A trailed disc harrows with 12 ft 10 in and 15 ft 9 in working widths had hydraulically folded wing sections.

Unlike spring-tine cultivators the frame of the spring-tine harrow runs at or near ground level. Spring-tine harrows were originally horse-drawn and later towed by a chain from a tractor drawbar, with the spring steel tines loosening the soil and the harrow frame levelling the surface. Mounted spring-tine harrows, usually with three or four separate sections, were either suspended by chains from the frame or attached to a ground-level frame.

3.9 The Stanhay Salo harrow was a modern version of the original wooden-beamed Dutch harrow.

Some farmers used Dutch harrows, or scrubbers, to prepare a level seedbed for drilling sugar beet. The first Dutch harrows, pulled with a length of chain by a horse or tractor, were little more than a number of heavy wooden slats held together by steel or wooden cross braces. Some tractor-mounted Dutch harrows had a levelling board at the front and rows of tines and a rear crumbler roller.

The chain harrow, mainly used for aerating grassland or spreading molehills and droppings, was not unlike a piece of large-scale chain-link fencing. Towed by a horse or tractor, some chain harrows had flat links while others were reversible with one side smooth and the other spiked for use on pastures to tear out dead grass.

3.10 John Wilder at Wallingford in Berkshire made the Pitch-Pole self-cleaning harrow.

Horse-drawn saddleback harrows, shaped like a horse's saddle, were used one row at a time to kill weeds on the top and sides of potato ridges. In more recent times chain and saddleback harrows have been suspended from a mounted harrow frame.

John Wilder at Wallingford made the first Wilder Pitchpole self-cleaning harrows in the 1930s. The trailed Pitchpole frame with two height-adjustable side wheels and a rear castor wheel had three or four rows of double-ended, square-section tines bolted at intervals on square shafts.

The 11 or 13 in long curved tines were locked in the working position by a cam arrangement controlled with a trip rope. With a quantity of trash collected on the tines, the trip rope was used to release the locking cams. This allowed the tines, aided by pairs of tumbler tines at right angles to the working tines, to rotate through 180 degrees and leave the accumulated trash on the ground. Various widths of Pitchpole harrows were made in the 1930s and 1940s. Mounted Pitchpole harrows were added in the early 1950s.

Power Harrows and Rotary Cultivators

The spring-tine cultivator was still the preferred tillage implement on many farms in the late 1950s but some farmers used a reciprocating bar power harrow to prepare their seedbeds. The principle of power cultivation dates back to the steam age and reappeared with the mighty Gyrotiller in the late 1920s.

The Horstman Cult-Harrow power harrow with two reciprocating tine bars was made in the late 1940s by Horstman Ltd at Bath and marketed by T Baker & Sons at Compton in Berkshire for tractors with a power take-off and hydraulic linkage. Sometimes mistaken for a Ferguson implement, the Cult-Harrow's reciprocating tine bars were belt-driven through a gearbox and to cranks on the tine bars. Complete with depth wheels the Cult-Harrow for the E27N Fordson Major cost £70.

An improved version of the Cult-Harrow with depth wheels for Fordson, David Brown and Nuffield tractors was made by Midland Industries at Wolverhampton in the mid-1950s. With the benefit of draught control

3.11 The tine bars on the late 1960s Benedict Weidner power harrow made 1,080 strokes per minute..

3.11A The two-bar Yeoman power harrow was imported by Colchester Tillage in the late 1960s.

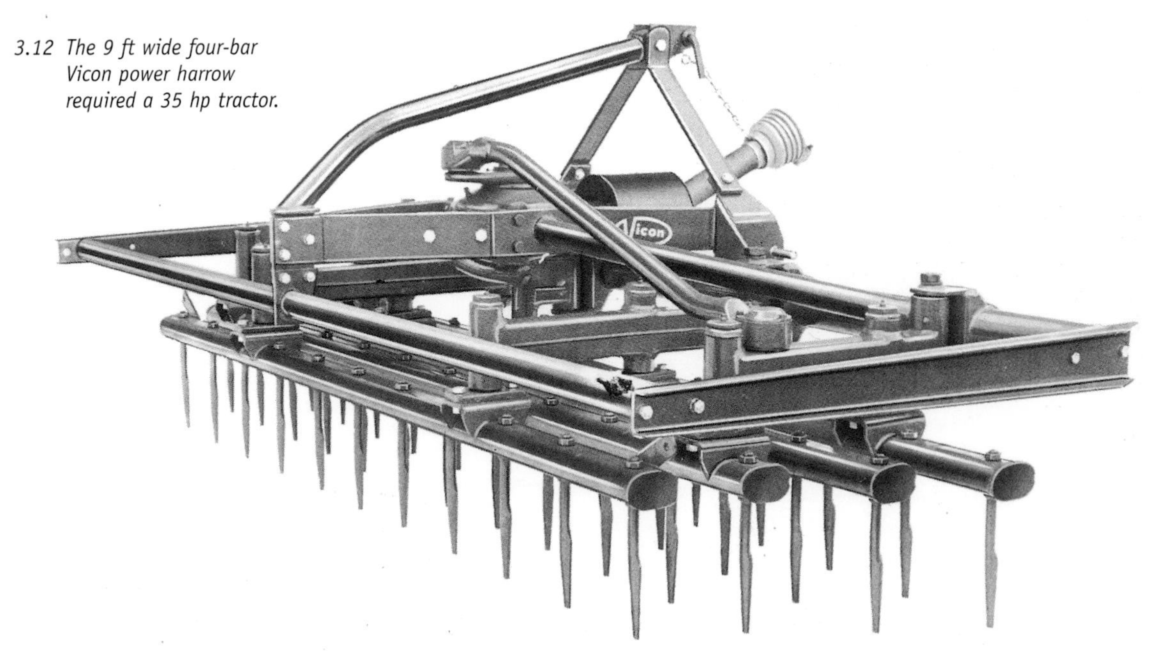

3.12 *The 9 ft wide four-bar Vicon power harrow required a 35 hp tractor.*

hydraulics the Cult-Harrow for Ferguson tractors didn't need depth wheels. However, the Horstman Cult-Harrow was a little ahead of its time and suffered teething problems with the tine-bar drive mechanism.

The advantages of reciprocating tine power harrows, which became popular in the early 1960s, were soon recognised, especially when preparing a spring seedbed on well-frosted soil. Early power harrows, including the Amazone, Benedict Soilmaster, Salopian, Vicon and Yeoman, were imported from continental Europe. The Soilmaster was made in working widths ranging from 6 ft 6 in to 16 ft. The three-bar Yeoman, imported by Colchester Tillage, was made with 8, 10 and 15 ft wide tine bars.

The drive mechanism on the Dutch-built Vicon four-bar reciprocating tine power harrow was arranged so that the length of stroke of each tine bar progressively increased from 4 in at the front to 20 in at the rear. The first Vicon power harrows were 9 ft wide, later models had working widths of 10 ft and 11 ft 6 in. Vicon two-bar power harrows were made in various widths in the late 1970s, the first being 8 ft wide. By the early 1980s there were 25 ft wide two-bar harrows for tractors of 150 hp or more which required folding wing sections for road transport.

Suitable for tractors of 40 hp or more, the 9 ft 10 in wide Salopian Huard Harrier power harrow, introduced by Salopian Kenneth Hudson in the early 1970s, had work rates of 3½–5 acres an hour.

Reciprocating bar and rotary power harrows were popular on arable farms in the early 1970s but some farmers preferred the alternative ground-driven rotary harrow with the tines driven by soil contact rather than by the more expensive tractor power shaft.

The Cee-Bee rotary harrow made by Parkin Engineering at Rotherham could be used at 10 mph. It had ten ground-driven rotors, each with eight curved vertical tines, giving a working width of 14 ft. Bamfords took over the marketing of the Cee-Bee harrow for a while but by 1974 it was back with Parkin Engineering when the ten-rotor model cost £330.

Other mounted ground-driven harrows used in the early 1970s included the Bomford Turbotiller, David Brown Albion rotary tiller and the Sampo rotary tiller. Made in 7 and 10 ft working widths, they had cross-shaped blades on either four or six angled shafts. Rotary tillers could be used for high-speed seedbed preparation, stubble breaking and minimal cultivation techniques but they did have a tendency to form mud balls in wet conditions.

The vertical-tined Lely Roterra rotary power harrow was introduced to British farmers 1968. The Roterra, protected by a number of patents, had the power harrow market to itself for several years.

3.13 A transport carriage with pneumatic tyred wheels was an optional extra for the widest 14 ft two-bar Vicon Power harrow.

Bigger and better models of Roterra appeared as the years passed and various designs of rear crumbler rollers became a standard attachment. Roterras with 5–10 ft working widths, two rotor speeds and a rear crumbler roll were made in the mid-1970s. By the early 1980s the biggest Roterra was almost 20 ft wide.

Howard, Krone, Kuhn, Lely, Rabe, Ransomes, Record, Rotomec, Ruston and SKH were some of the many makes of power harrow at the 1985 Royal Smithfield Show. Badge engineering disguised the

3.14 The Cee-Bee ground- drive rotary harrow was made by Parkin Engineering at Rotherham.

3.15 The four- and six-axle Bomford Turbotillers were used at speeds of up to 10 mph.

fact that many of them were of Italian origin. Ransomes introduced 8, 10 and 11 ft 6 in models of the Italian Pegoraro power harrow at the show and as Agrolux Ransomes they added more rugged 10–13 ft wide harrows for 75–170 hp tractors in 1988.

The 13 ft wide Ransomes power harrow was driven by a 1,000 rpm power take-off shaft and, in common with other makes, optional extras included rear crumbler or packer rollers and a mechanical or hydraulic linkage for grain drills.

3.16 The first Lely Roterra rotary power harrows were sold to British farmers in 1968.

3.17 A packer roller or open cage roller could be used with a Ransomes power harrow.

The Maschio Recotiller imported by Ruston's Engineering and the Howard Farmhand Harrovator with a three-speed gearbox were among the makes of rotary power harrows on the market in the mid-1980s.

The rotary cultivator, with the Rotavator the best-known example, was used on many farms to prepare seedbeds for the potato crop. AC Howard invented the Rotavator in Australia in the 1920s and his name remains linked with rotary cultivation to the present day. Captain EN Griffiths, who had imported a Howard Rotavator to use on his farm, obtained a licence in 1931 for J&F Howard Ltd of Bedford, a company unrelated to AC Howard, to manufacture the Howard Rotavator.

Captain Griffiths and AC Howard formed Rotary Hoes Ltd in 1938 at East Horndon, later renamed West Horndon, where pedestrian-controlled and tractor-powered Rotavators were made for the next thirty years or so.

Rotary Hoes also introduced the Rotehoe agricultural drainage machine in 1942 and the Dungledozer farmyard manure loader in 1944. Both were built around a Fordson tractor. Dungledozer production was limited to sixteen and at least half of them were sold abroad.

Tractor-mounted Rotavators included the offset 40 and 50 in E Series made for the most popular tractors of the day. The 130 in wide model M Rotavator was suitable for high-powered wheeled and crawler tractors. Pan busting and rotary cultivation were combined with the launch of the Howard Underbuster, a Rotavator with subsoiling tines attached to the frame. The Howard Rotaseeder and Rotacaster with a drill unit mounted above a Rotavator were used for direct drilling. Howards also produced a Rotavator with a spiked rotor for minimal cultivation work.

Rotary cultivators were rather slow, the 50 in E Series only achieved an output of 2½ acres an hour at 5 mph and this was only possible in the lightest soils. At a more sedate 2 mph in heavier soil the output was little more than an acre an hour. Work rates were considerably improved by the arrival of more powerful tractors and the Selectatilth gearbox with a choice of six rotor speeds.

3.18 *The Selectatilth gearbox on the Howard E Series Rotavator provided six rotor speeds ranging from 125 to 225 rpm.*

The Howard Rotavator Co moved to new factories in East Anglia in the early 1970s where a wider product range, including Rotaspreader side-spread manure spreaders, balers, mowers and grain drills for Massey Ferguson, was made until the company ceased trading in 1985.

Dowdeswell Engineering bought the Howard factory at Harleston in Norfolk and continued production of pedestrian-controlled rotary cultivators and manure spreaders under the Dowdeswell name. Farmhand acquired the Howard name and the Howard factories in continental Europe where they made tractor-mounted Rotavators, at first under the Farmhand Howard name but later with a Howard GB badge.

The Bonser Agrotiller, Kronevator and Landmaster rotary tillers, Lely Buryvator and the Martin-Markham

3.19 *Frank Bonser made trailed and mounted Agrotiller rotary cultivators in the early 1960s.*

3.20 *Sales literature explained that the Lely Buryvator, which carried the soil and trash over the top of the cultivating rotor, was the only rotary cultivator that buried the vegetation in one run, leaving a seedbed ready to be drilled.*

Powertiller were among a lengthy list of mainly mounted rotary cultivators in competition with the Howard Rotavator in the 1960s and 1970s. The trailed Bonser Agrotiller rotary cultivator with a 60 or 70 in rotor was made for the David Brown 900, Massey Ferguson 65 and other popular tractors of the day.

3.21 *Optional equipment for the Bomford Dyna-Drive ground-driven rotary cultivator, introduced in 1982, included a packer or a crumbler roller and a linkage for a rear-mounted seed drill.*

Rotary cultivators cut the soil and threw it against the hinged shield at the rear to shatter the clods. The Lely Buryvator blades, which rotated in the opposite direction to the blades on other rotary cultivators, carried the soil over the top of the rotor where a screen separated trash from the soil. The separated trash was returned to the ground and buried by the soil.

Subsoilers

Single- and double-leg trailed subsoilers and subsoiling tines bolted to or replacing the rear body of a trailed plough were rarely used in the 1950s and 1960s. However, in order to overcome soil compaction and drainage problems caused by the ever-increasing weight of tractors and the cumulative effect of minimal cultivations and direct drilling techniques, subsoiling became almost a necessity in the early 1970s.

Ransomes, Ferguson and a few other companies had made subsoilers for a number of years. Then, with the increased interest in subsoiling, Bomford & Evershed, Brown, Cooper, Howard, McConnel and others entered this market.

Simple single- and twin-leg subsoilers were not very effective at breaking up heavily panned soil and more sophisticated subsoilers were developed to solve the problem. The McConnel Shakaerator, introduced in 1976, was still made in the early 1990s. Designed in Australia, the Shakaerator was a hybrid of a subsoiler and a chisel plough with a pto-driven vibrator unit on the main frame. This vibrated the entire implement, moving it up and down and sideways several times a second.

The American-designed Avery vibrator kit, first marketed by a Norfolk farmer in 1977, could be fitted

3.22 There was a rack-and-pinion lifting mechanism on both wheels of Ransomes trailed subsoilers.

3.23 A pto-driven vibrating mechanism oscillated the tines on the McConnel Shakaerator heavy-duty cultivator and subsoiler.

3.24 The twin-leg Cooper subsoiler had wing shares on the subsoiling legs.

3.25 A pto-driven oscillating blade was attached to the front of the Lely Brenig subsoiling leg.

3.26 The Flatford Progressive Soil Ameliorator with a 13 ft maximum working width had seven sweep tines, two cutter tines, seven chisel tines and two winged subsoiling tines at the back. Sales literature in 1979 suggested that with a full set of tines the Ameliorator would require a tractor in the 180 to 200 hp bracket.

3.27 A 90 hp two-wheel drive tractor or a 60 hp crawler was recommended when working at a depth of up to 20 in with the three-bladed Howard Paratyne subsoiler.

to most makes of chisel plough, subsoiler or cultivator. The unit consisted of two power-driven contra-rotating weights running at 40 rpm which vibrated the tines as the subsoiler was pulled through the soil. Prospective purchasers were told that because of the vibrating action of the rotating weights it was important to tighten all nuts and bolts at regular intervals. Drivers were also advised to stop the vibrator unit before lifting the implement from work.

Satisfied that vibrating tines increased the shattering effect on the soil, Lely introduced the Brenig subsoiler with a power-driven, vertical-front oscillating tine in front of the main leg in 1977.

By the late 1970s it was well established that fitting wings to the subsoiler share increased the shattering effect on the soil. Wings were much cheaper than power-driven vibrators and most subsoiler manufacturers added wings to their list of optional extras.

Multi-leg subsoilers, including the Bomford Earthquake, Haylock four-leg trailed model, McConnel Commando, Flatlift, Howard Paraplow, Cooper three-leg Topsoiler and the Ransomes seven-tine Subtiller, were all designed for 150–250 hp tractors. They dramatically extended the range of pan-busting equipment in the early 1980s.

However, within a few years the mouldboard plough returned to favour and when most of the 250 hp-plus two-wheel drive tractors disappeared from the scene many farmers saved both time and money by leaving the subsoiler in the barn.

Combination Cultivators

Some manufacturers used the multi-leg subsoiler as the basis of multi-function cultivators. Large mounted or trailed tool frames, with different combinations of shallow and deep working tines, discs and crumbler rollers, were developed to carry out a number of soil treatments in a single pass.

The aptly named Progressive Soil Ameliorator (the dictionary defines ameliorate as to improve or make better) was a heavy-duty example of a one-pass cultivator with chisel plough-type tines at the front and deeper working subsoiling tines at the rear.

Lighter one-pass tillage machines with various combinations of rigid or spring tines, discs, levelling bars and crumbler rolls were also made in the 1980s. The trailed Taskers Tillage train was one example of this type of multi-purpose tillage equipment. It had two rows of cultivator tines followed by two gangs of disc harrows and a pair of transport wheels raised and lowered with a hydraulic ram. It could be used at 10 mph. The 8 ft model needed a 100 hp tractor, and 180 hp was required for the 12 ft Tillage Train.

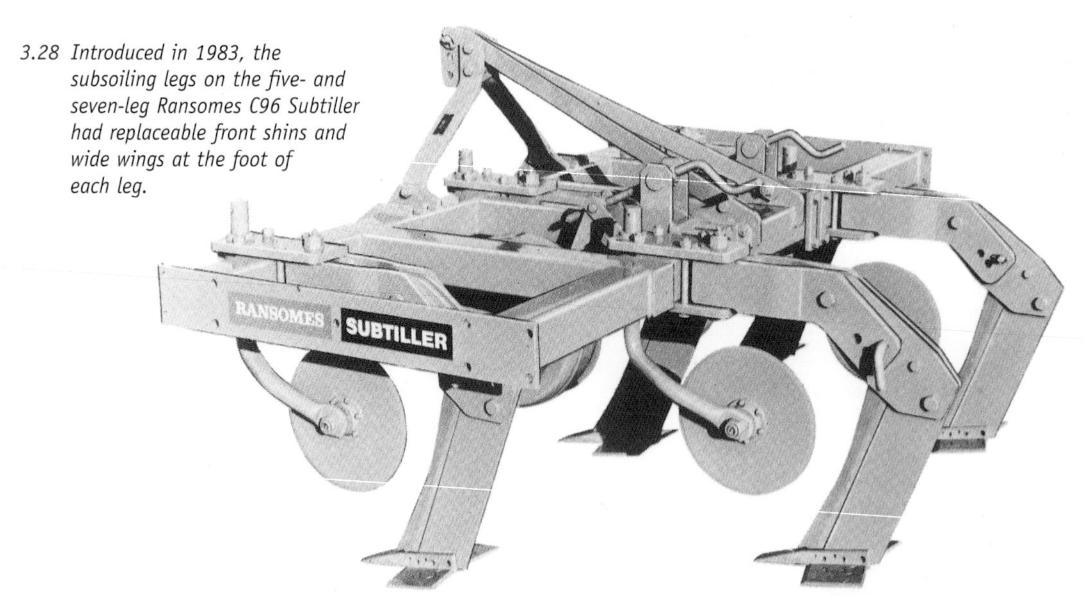

3.28 Introduced in 1983, the subsoiling legs on the five- and seven-leg Ransomes C96 Subtiller had replaceable front shins and wide wings at the foot of each leg.

3.29 The Taskers Tillage Train had two rows of cultivator tines followed by two gangs of disc harrows and hydraulically raised and lowered transport wheels. A 100 hp tractor was needed to pull the 8 ft wide Tillage Train at speeds of up to 10 mph.

3.30 The Kongskilde Germinator levelled, cultivated and consolidated the soil in one pass.

3.31 Bernard Krone marketed the Becker combination cultivator in the UK.

Steerage Hoes

Jethro Tull, who promoted the idea of drilling crops in rows rather than broadcasting seed to the four winds in the mid-1700s, also introduced the horse hoe. Various shapes and sizes of drill and horse hoe were made by blacksmiths. By the late 1940s some were being converted for use with a tractor, either towed from a tractor drawbar or hitched to the three-point linkage.

The Edlington steerage horse hoe made at the Phoenix Iron Works at Gainsborough was typical of those made in the 1930s and 1940s. The basic hoe for sugar beet cost £20 in 1948 and an alternative model with corn hoes to follow a thirteen-row corn drill was £23 15s 0d. The hoe blades could be moved on the frame to suit different row widths and could be set close up to the rows of young plants. The frame and hoes were suspended with a chain for transport. An advertisement suggested that the Edlington hoe was so easy to handle that a boy could use it without any trouble.

3.32 *The Ferguson rear-mounted steerage hoe with independently mounted hoes was first made in the early 1950s.*

Several makes of front-, mid- and rear-mounted hoes appeared on the market in the 1940s. One-man front- and mid-mounted hoes were more economical in labour costs but any lapse in concentration by the driver, especially when using a front-mounted hoe, was likely to hoe out both weeds and crop.

The two man rear steerage hoe was more accurate. Because it left the narrowest possible width of undisturbed soil on each side of the row at singling time it was popular with the hand hoers who would be able to chop out more rows of beet in a day.

Tractor steerage hoes did not really catch on until the arrival of the hydraulic three-point linkage in the mid-1940s. However, WT Teagle at Blackwater in Cornwall introduced what was said to be the world's first steerage hoe in 1942. Mounted on the back of a Fordson Standard tractor it was steered by a single rear castor wheel controlled by a man standing on a platform.

To satisfy the demand for the Teagle hoe it was also made by Boulton & Paul. Ransomes, which was making horse hoes in the late 1800s, introduced a rear-mounted hand-lift toolbar with cultivator tines and hoe blades in

3.33 *Hoeing sugar beet with an Allis-Chalmers ED 40 and mid-mounted hoe.*

3.34 The mid-1950s Standen Crop Hopper could be used to hoe, spray and cultivate at the same time.

the mid-1930s for use with Farmall, Case, Fordson, John Deere and Allis-Chalmers tractors. The toolbar with nine hoe blades cost £24 5s 0d and a chain-lift mechanism, shaft driven from the tractor transmission, was available at extra cost.

The Ransomes C60 rear-mounted toolbar with a steerage hoe and other attachments was made for the E27N Fordson Major. The hoe was steered by a tiller arm, linked to the lower lift arms, from a seat mounted above a single pneumatic-tyred rear wheel.

There were plenty of steerage hoes on the market in the late 1940s and early 1950s. Several companies, including Acrow-Pygro, Belton Bros & Drury, Ferguson, Johnsons, Russell, Stanhay and Standen, made rear-mounted hoes. Gloster, Nicholson, Standen, Stanhay and others built front- and mid-mounted hoes.

The rear-mounted Acrow-Pygro tractor hoe, for inter-row hoeing of rowcrops and cereals, had individually mounted hoe units on a toolbar with a pair of rear castor wheels. The operator, seated at one side of the hoe, used a steering wheel to keep the hoe units on the rows. Sales literature explained that the articulated hitch meant that no matter whether the rows were straight or wavy, the steering was sensitive enough for the operator to steer with absolute accuracy and no plants would be spoiled.

The four-, five- or six-row Belton beet hoe made by Belton Bros & Drury in Lincolnshire had individual depth control on each hoe unit.

The rear-mounted Whitlock steerage hoe, introduced in 1956, was a one-man outfit achieved by driving the tractor in reverse with the driver seated somewhat precariously on the tractor bonnet. Suitable for Ferguson and Fordson tractors, the Whitlock hoe's price ranged from £69 to £127 2s 6d depending on the number of rows and type of hoe blades.

The four-, five- and six-row Standen Hereford hoe was a typical 1950s one-man, front-mounted steerage hoe. The hoe units were independently mounted and sprung to follow the ground contours. The toolbar was raised and lowered with a pulley and cable arrangement operated by the hydraulic linkage.

Some farmers used a Bean self-propelled toolbar with independently sprung hoe units made by Humberside Agricultural Products. Others used one of the small three- or four-wheel market garden tractors such as the David Brown 2D, Gunsmith and Newman or a Ransomes MG crawler with front-mounted hoes. Even so, at a spring sugar beet demonstration held in Suffolk in 1950 a horse hoe with a horse between the shafts was to be seen working in the same field alongside the latest tractor steerage hoes.

3.35 The Ransomes FR Cleanrow chemical hoe could apply herbicides at a rate of 25 gallons per acre.

The late 1960s Ransomes FR Cropguard Cleanrow inter-row chemical hoe with Vibrajet no-drift nozzles provided an alternative method of weed control in rowcrops. Independent floating steel crop guards were used to keep the chemical, pumped to the nozzles from a 30 or 60 gallon front-mounted tank, off the rows. The 9 ft wide Cleanrow chemical hoe could be extended to a maximum width of 12 ft and it was possible to treat between five and twelve rows in a single pass.

Weeders

Weeders were mainly used to kill off small weeds in emerging potato or cereal crops and sometimes to stir up the weeds between rows of sugar beet. Small weeder units made by Reekie Engineering at Arbroath were used with a mounted cultivator. As the cultivator tines worked the soil between the rows of potatoes, the weeder units stirred up the weeds growing on the ridge.

3.36 Cultivating potatoes with a Ferguson cultivator and Reekie weeders.

3.37 *It was claimed that this late 1950s Massey Ferguson 71 tine weeder could weed up to 60 acres in a day.*

Aldersley, Ferguson and others made full-width weeders with two or three rows of spring tines for use in cereal and pea crops. Trailed and mounted Aldersley Universal weeders were made in 12 ft and 14 ft 6 in widths, with the wing sections folded for transport. Trailed weeders had a number of tine sections suspended from an implement frame and were made with a drawbar or horse shafts and a pair of wheels at the back.

The introduction of selective weedkillers made the tined weeders obsolete but in more recent times environmental considerations have seen Lely and a few other companies re-introduce tined weeders to the British market.

3.38 *In the 1990s the Lely weeder provided an environmentally friendly way to kill weeds in the 1990s.*

Chapter 4

Grain Drills

In 1679 John Worlidge came up with the idea of using a machine to plant seeds in rows, thus making it easier for hand labour to remove weeds from growing crops. Fifty years later Jethro Tull realised that there were advantages to be gained from this and in 1731 he showed the design of his own seed drill to local farmers. Two years later Tull published the book *Horse Hoeing Husbandry* in which he set out in some detail the benefits of rowcrop farming.

Turnip Townshend tried out the idea on his Norfolk estate where he used hand labour to plant seed in rows but was unable to find a blacksmith and wheelwright able to make Tull's drill design.

The Rev James Cooke introduced a modified version of Tull's drill in 1782 and in 1800 James Smyth, a blacksmith at Peasenhall in Suffolk, made the first Smyth cup-feed drill. Smyth Non Pareil cup-feed drills with Suffolk coulters were made at Peasenhall for the next 160 years. An entry in *Kelly's Directory* for Suffolk in 1883 recorded that James Smyth & Sons had extensive implement works at Peasenhall with other premises at Witham in Essex and rue Lafayette in Paris.

By the mid-1800s, in order to meet local demand in other parts of the country, agricultural engineers, including Yates of Doncaster, LR Knapp, Richard Hornsby, Richard Garrett and James Coultas, were making drills based on the original Cooke design. A report published in 1860 noted that a farmer who used a Hornsby cup-feed drill, which cost £37 10s, could expect to make a profit of £1 15s 8d for an acre of wheat. However, this profit depended on the crop having a liberal dressing of 100 lb of general fertiliser at a cost of 8s 4d an acre.

The seed barrel was at the back of a cup-feed drill so that the horseman or his assistant, walking behind the drill, could start and stop the drive to the seed barrel and keep an eye on the flow of seed from the cups. Even with tractor-drawn drills it was still necessary to walk behind until someone thought of providing a riding platform. Two seed barrels were supplied with the drill with each barrel having two sizes of seed cup, making the drill suitable for sowing all sorts of seed from barley to beans and clover to cabbages. The seed rate was changed by using different-sized driving gears to change the speed of the seed barrel and occasionally by using different-sized cups.

The coulters were spaced 7 in apart and seed could be drilled at different row spacings by shutting off the flow of the seed to selected coulters. Some farmers filled alternate seed cups with putty or even chewing gum in order to achieve a degree of plant spacing in the rows and reduce the time taken to hand single root crops.

4.1 The horses were hitched to a steerable fore-carriage on early Smyth cup- feed drills.

Horse-drawn cup-feed drills enjoyed the lion's share of the market in the 1920s and 1930s but by 1940 they were in competition with the internal force-feed drill made in America by International Harvester, Oliver, Massey-Harris and others.

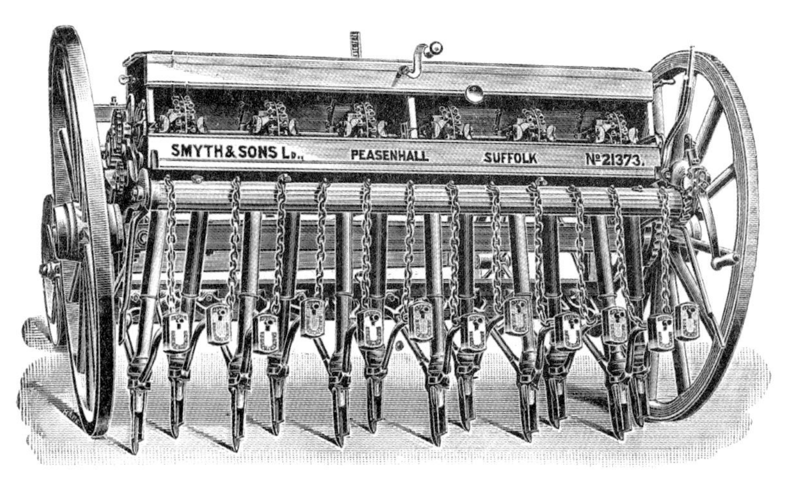

4.2 Nine- to sixteen-row Smyth cup-feed drills were made in the 1940s and 1950s.

The internal force-feed mechanism had a number of rimmed seeding discs, one for each coulter, turning in the bottom of the seed hopper. Each disc had a wide rim for oats and peas and a narrower rim for wheat and barley; a cover plate was placed over the unwanted side of the disc. Serrations on the inside of the disc rims carried seed to the seed tubes which delivered them to the coulters. Seed rate was adjusted by using different sets of gears or chain sprockets from the land wheels to vary the speed of the shaft carrying the seed discs.

The first horse-drawn International Harvester internal force-feed grain drills were made in America in 1902 and tractor-drawn International combined grain and fertiliser drills appeared in the UK in the early 1930s. Massey-Harris, which was associated with Blackstone at Stamford at the time, were making eleven- and thirteen-row manual and self-lift grain and combine drills for British farmers. With few exceptions the seed tubes on combine drills supplied both the seed and fertiliser to the coulters.

Smyth, Knapp, Yates and others were still making cup-feed drills in the mid-1940s when Knapp also made up to seventeen-row hand-lift horse-drawn and self-lift tractor-drawn Monarch internal force feed drills.

British-made Albion and Dening Somerset internal force-feed drills, together with International Harvester, Massey-Harris and Oliver drills from America, were popular in the mid-1940s. There was a choice of Suffolk (fen), hoe, or disc coulters for the thirteen, fourteen and fifteen-row Albion drills made by Harrison, McGregor & Guest at Leigh in Lancashire.

Optional equipment for Albion grain and grass seed drills included a horse pole and whippletrees, a steering fore-carriage or a tractor hitch, a foot board and harrow drawbar. There was reduction gearing for small seeds and even a seat for the man whose job was to walk behind the drill.

The steel-framed Dening Somerset twelve-row manual lift internal force-feed combine drill made at Chard had a star wheel fertiliser feed and disc and Suffolk or hoe coulters. There was a choice of steel wheels or

4.3 The Massey-Harris 720A grain and fertiliser drill had a land wheel-operated lift mechanism.

pneumatic tyres and two sets of gears for changing seed and fertiliser application rates were carried on the drill frame. Later models of Dening external force-feed twelve- and sixteen-row self-lift tractor drills had flexible 'non-kinking' metal grain tubes and rubber fertiliser tubes.

4.4 *Pierce Victor grain drills were made in Ireland.*

Internal force-feed drills imported from America in the 1940s included those made by Massey-Harris and International Harvester and the thirteen- and fifteen-row power-lift Oliver drills sold in the UK by John Wallace at Glasgow.

A number of companies, including Wrekin and International Harvester, also made fluted roller external force-feed drills in the mid-1940s. The fluted roller feed mechanism, suitable for drilling a much wider range of crops, was used for the Ferguson Universal grain drill introduced in 1950 and the Ferguson Universal grain and fertiliser drill added in 1951.

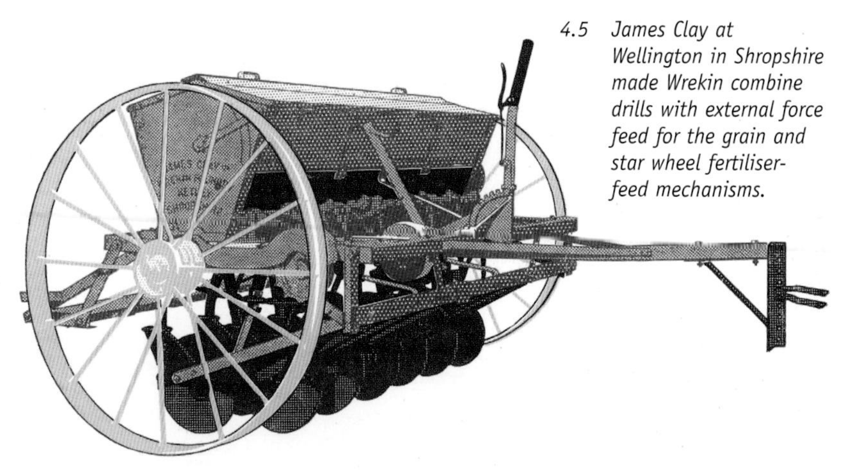

4.5 *James Clay at Wellington in Shropshire made Wrekin combine drills with external force feed for the grain and star wheel fertiliser-feed mechanisms.*

The new thirteen-row Ferguson grain drill, advertised as one drill to do the work of three, made obsolete the need to use different drills for grain, grass and root crops. The Ferguson drill was hitched to the tractor three-point linkage and the driver was able to raise and lower the coulters, engage and disengage the drive to the feed mechanism and check the flow of seed from the driving seat.

The fertiliser hopper with its star wheel feed mechanism was lowered to a horizontal position for cleaning and small root seed hoppers could be placed at the same intervals as the row spacings across the width of the main grain hopper.

Like most grain drills at the time, the Wrekin external force-feed drill, made by James Clay at Wellington in Shropshire, could have disc or Suffolk coulters at 7 in row spacings. The drill had a steel grain hopper but a non-corrosive wooden hopper was used for the star wheel fertiliser feed mechanism. A tractor drawbar was standard but Wrekin combine drills were still available with a horse pole.

Philip Pierce & Co of Wexford made eleven- and thirteen-row internal force-feed corn drills in the early 1950s when Massey-Harris introduced the 720A grain and fertiliser drill and the Sunshine Suntyne cultivator drill. The Sunshine Suntyne and the Sundrill grain and fertiliser and plain grain drills were made by the Butterley Co in Derbyshire in the mid-1950s and

4.6 *The Ferguson Universal drill made the need to have separate drills for different crops obsolete.*

marketed by the Sunshine Harvester Co in London.

The thirteen-, fifteen- and twenty-row 720A combine drill had steel-covered wooden hoppers for the internal force feed and the star wheel fertiliser mechanisms, a land wheel-operated power lift clutch and the choice of disc, hoe or Suffolk coulters. An optional transporter truck was made for the twenty-row 720A drill. Sales literature explained that the secret of the accuracy of the 720A was its effective force-feed sowing mechanism that distinguished it from ordinary drills.

Massey-Harris-Ferguson introduced the 500 Series drills to British farmers in 1955. Described as a modern three-

4.7 *The first Bamford Octopus seed drills were made in 1967.*

in-one drill which carried out final cultivations, killed weeds and planted the seed in one pass, the twelve-row drill with twenty-five spring tines and sixteen-row drill with thirty-three tines had respective sowing widths of 7 ft and 9 ft 4 in.

In 1957 Bamfords of Uttoxeter introduced the thirteen-row SF13 drill which, because of its looks, became known as the Bamford Octopus. Designed in Norway, the SF13 had a feed mechanism totally new to UK farmers. The land wheel-driven feed mechanism consisted of a stainless-steel cone which, depending on forward speed, revolved at speeds between 300 and 1,200 rpm. Seed was thrown centrifugally through an adjustable outlet in the cone into a manifold from where it was distributed through a series of tubes to the thirteen coulters. A grain and fertiliser version of the Octopus had a second stainless-steel cone-shaped hopper inside the grain hopper which rotated at similar speeds to supply fertiliser into a separate manifold and delivery tubes.

Mounted fifteen- and nineteen-row and semi-mounted thirty- and thirty-eight-row Octopus drills were made in the mid-1960s. The S30 and S38 were twin thirty- and thirty-eight-row semi-mounted drills for tractors with category II hydraulic linkage. The drill, mounted on rear castor wheels, was in two parts with each section having a cone-shaped seed hopper and either fifteen or nineteen coulters.

The Bamford Octopus nineteen-row KO 19 launched in 1967 had a larger hopper and a two-speed vee-belt pulley to drive the seed mechanism. The lower belt speed was used to sow grass seeds at speeds of up to 5 mph.

Some 160 years after the first Smyth drills were made in Suffolk they were given a makeover. In 1962 the seed hopper was turned round with the cup-feed mechanism in full view from the tractor seat. The new Model 22 Smyth drill with the option of steel- or pneumatic-tyred wheels had an 8 ft 3 in sowing width and the coulters at 4¾ in spacings were raised and lowered with a hydraulic ram.

Unlevel ground affected the efficiency of cup-feed drills and when used at speed some of the seed fell from the cups back into the hopper. The acquisition of the Smyth business by Johnson's Engineering in the mid-1960s signalled the end of the traditional Smyth cup-feed drill. After closing the Peasenhall works Johnson-Smyth introduced the fifteen-row combine drill in 1967.

4.8 The Bamford Octopus S30 grain drill with a 17 ft sowing width was made for tractors with category 2 hydraulic linkage.

The new drill with external force-feed mechanisms in both hoppers was made at March in Cambridgeshire. The Johnson-Smyth drill had Suffolk coulters, pneumatic tyres, seed level windows in the grain hopper and a flashing light to warn the driver when the fertiliser hopper was almost empty.

A range of Nordsten narrow spacing drills from Denmark, originally imported by Shermore Ltd at Norwich in the mid-1960s, included the 29 and 33 row mechanical-lift trailed Ceres drills and 21 to 50 row fully mounted Lift-o-Matic drills.

4.9 A hydraulic ram was used to lift the Smyth Model 22 cup-feed drill coulters out of work.

Ransomes, which was making cultivator drills in the 1900s, resumed interest in the grain drill market with the Johnson-Smyth combine drill when they bought Johnson's Engineering at March. Shortly after the 1967 takeover Ransomes added six models of studded roller feed Nordsten Lift-o-Matic grain and combine drills to their product range.

4.10 Ransomes made cultivator drills in the early 1900s; this one has a grass seed box for under-sowing grass seed while drilling a cereal crop.

The feed rollers were driven from a sixty-speed gearbox and the coulters were spaced at 4 in intervals across the 8 ft 3 in–19 ft 9 in sowing widths. Nordsten Combi-Matic grain and fertiliser drills with separate coulters placing fertiliser between each pair of seed coulters were added in 1971. A new 10 ft Nordsten cultivator drill made its debut in 1976.

4.11 *Ransomes introduced the Nordsten Lift-o-Matic grain drill in 1967.*

Most drills sold in the late 1950s, including those made by Allis-Chalmers, Lundell, Wexford and Wrekin, were combined grain and fertiliser drills. However, within a few years some farmers left their corrosion-prone combine drills in the barn and used narrow spacing grain drills with 3½–4½ in coulter spacings. The Nordsten Lift-o-Matic, one of the first to appear in the UK, was soon followed by several more mainly mounted drills including the Fiona, McCormick International, Stegsted and Viking from Scandinavia and the Europlex from Belgium. Some could also be used to drill at the traditional 7 in row spacing but drilling in narrow rows at the same seed rate was deemed to give the young plants more room to grow.

The Swedish McCormick International S6-1 plain grain drill could, for example, be used with the coulters at 5, 6 or 7 in spacings and the sixty-speed gearbox provided a wide range of seed rates. Of lighter construction than earlier grain drills, the lightweight Suffolk coulters were likely to bend if the tractor driver reversed the drill with the coulters in the ground.

4.12 *McCormick International S6-1 plain grain drills were made in Sweden.*

Cultivator Drills

4.13 *Carier cultivator drills were made at Braintree in Essex.*

Following an earlier trend to use chisel ploughs and heavy cultivators rather than mouldboard ploughs, the mid-1960s saw a move towards the cultivator drill. Cultivator drills were not a new idea as they were already in widespread use in Australia in the early 1950s when a few British farmers drilled their cereal crops with a twelve- or sixteen-row Australian-built Massey-Harris Suntyne cultivator drill.

The Carier cultivator drill, introduced in 1964, with the seed hopper mounted above a spring-tine cultivator, was one of the first of a new generation of cultivator drills. Several cultivator drills, including those made by Ben Burgess, International Harvester, Fiona and Jones among others, were popular in the early 1970s but direct drills used on unbroken stubble were also coming into use.

The fourteen- and twenty-row Fiona cultivator drills with 6 in row spacings had studded roller feed and the seed rate was varied through a sixty-speed gearbox. Sales literature explained that the drill, normally used on ploughed land, was used to cultivate the soil and place fertiliser 3½ in deep on the first pass. Then, after refilling the hopper with seed, further seedbed preparation and sowing the seed was completed on the second pass of the drill.

The 7 in row spacing International Harvester 6-2 cultivator drill from Australia, with seed and fertiliser hoppers, was suitable for use both on ploughed land and for direct drilling. The thirty-three cultivator tines on the sixteen-row drill and forty tines on the twenty-row model were arranged in three staggered rows across the width of the drill.

Unlike the Carier, Fiona and International Harvester drills the Ben Burgess Cultor-Seeder with an 8 ft 9 in sowing width and land wheel-driven seeding mechanism did not have seed tubes or coulters. The seed and fertiliser were broadcast immediately behind a front levelling board and spring tines across the full width of the drill. The seed was covered by a rear levelling board and a row of thin flexible tines.

The Jones 519 cultivator drill, made in the early 1970s by Jones Balers at Mold, had work rates of up to 6 acres an hour. It consisted of a Bamford Octopus centrifugal seed drill mounted above a nineteen-tine

4.14 *The Burgess Cultor-Seeder was used to broadcast seed and fertiliser on ploughed land.*

Kongskilde Triple K spring-tine cultivator. The coulters could be set to drill at 5–6½ in row spacings.

Accord and other makes of pneumatic feed seed drill were coming into use on some farms in the early 1970s. The first Accord pneumatic drill, which appeared in 1966, had a land wheel-driven central seed metering mechanism and a pto-driven fan. Four models of the Accord pneumatic seed drill were marketed by Alpha-Accord at Ampthill in Bedfordshire in 1970. The smallest was a twenty-nine-row drill with 7 in coulter spacings. The widest 23 ft Accord drill with sixty-four coulters spaced 4½ in apart had a 12 cwt seed hopper and an end-tow arrangement for transport.

The mid-1970s fifty-row Bamlet Tive Sow-Jet pneumatic feed drill, with a 3 ton capacity hopper, was suitable for all sorts of seed from grass to peas. The coulters were spaced 4¾ in apart but by blanking off certain feed rollers the Sow-Jet could be used to drill at 9½ and 19 in row spacings.

Direct Drilling and Minimal Cultivations

Direct drilling was a serious alternative to minimal tillage and traditional seedbed cultivations by the mid-1970s. Direct drills worked best on stubble fields where the straw had been baled or burnt to leave a relatively trash-free surface. Farmers using this technique sprayed stubble fields with Gramoxone to kill off the weeds before sowing the next crop with either a heavy-duty disc coulter direct drill or a power harrow and drill combination.

The Rotadrill and the later Rotacaster minimal cultivation drills were made by Howard Rotavators. Introduced in 1967 the Howard Rotadrill had a Nodet land wheel-driven feed mechanism in a seed hopper mounted above a 70 or 80 in E Series Rotavator. Independently sprung Suffolk coulters at 6 in spacings drilled seed into freshly Rotavated soil.

The Rotacaster, which superseded the Rotadrill in

4.15 The Howard Rotacaster could be used to either broadcast or drill cereal crops.

4.16 The Howard Rotaseeder drilled seed in narrow slots cut in unploughed land.

1971, was similar with a drill box mounted above an 80 or 90 in wide E Series Rotavator but it was more versatile than the Rotadrill. The Rotacaster could be used to broadcast seed from the thirteen tubes suspended in front of the cultivating rotor or drill the seed in rows with the seed tubes in coulters spaced 6 in apart at the rear.

4.17 *The minimal cultivations Kronevator and seed drill combined seedbed preparation and sowing in one pass.*

4.18 *Disc, Suffolk and hoe coulters were available for the Massey Ferguson No. 29 combine drill.*

The Howard Rotaseeder, also introduced in 1967, with a Nodet external force-feed drill mounted above a Rotavator, was another approach to direct drilling. Fifteen sets of narrow Rotavator blades cut slots 5 in apart in unbroken ground and seed was placed in these slots by narrow coulters. Howard Rotavators made the Rotacaster and Rotaseeder direct drills until 1973 when the remaining stock of parts was sold to the West Midlands Farmers Association at Gloucester.

The Kronedrill, made by Bernard Krone in Germany, was similar to the Howard Rotacaster. Suitable for 65–75 hp tractors, the Kronedrill was a combination of an 80 in wide Kronevator rotary cultivator and an Amazone seed drill. It was used to broadcast seed in front of the cultivating rotor or to drill seed with fourteen coulters at the back of the rotary cultivator. The drill-feed mechanism was land-wheel driven; a rear crumbler roller was an optional extra.

Direct drills, combine and plain grain drills were made throughout the 1970s. The Massey Ferguson 29 combine drill with an internal force-feed mechanism for the grain and star wheel fertiliser feed was one of the more popular models. The improved Massey Ferguson 30 with an external fluted roller feed for seed and fertiliser replaced the earlier model in 1974 and, like its predecessor, had thirteen, fifteen or twenty disc coulters or Suffolk coulters at 7 in row spacings. The partition between the grain and fertiliser hoppers could be reversed to provide large grain and smaller fertiliser compartments or vice-versa.

Other grain drills in the mid-1970s included the trailed and semi-mounted International 510 with disc or hoe coulters and the narrow spacing twenty-one-row Allis-Chalmers 214 marketed by Jones Balers at Mold. Nordsten drills, imported by Ransomes included the Lift-o-Matic 400 narrow spacing grain drills with 8 ft 3–19 ft 8 in working widths and two models of Nordsten grain and fertiliser drills. Separate grain and fertiliser coulters with the fertiliser placed at one side of the rows of grain were a feature of the 9 ft 9 in and 13 ft 3 in wide Farmhand combine drills.

4.19 Bettinson 3D direct drills had triple-disc coulters.

4.20 The Massey-Ferguson 130 direct drill had a reversible partition between the seed and fertiliser hoppers.

Several new direct drills, including those made by Barford, Bettinson, Craven Tasker, International Harvester, Massey Ferguson and Moore, appeared in the mid-1970s. Most direct drills had triple disc or spring-tine coulters and plenty of built-in weight to assist coulter penetration in hard-baked ground. The Barford GF 150, Bettinson 3D and Massey Ferguson 130 combine drills with big seed and fertiliser hoppers had fifteen sets of triple disc coulters at 7 in row spacings.

The Massey Ferguson 130 coulters were mounted on parallel linkage frames and, for added coulter penetration, wheel weights were used and the drill wheel tyres could be filled with up to 440 lb of water ballast. Optional concrete ballast weights for even more coulter penetration were made by Massey Ferguson.

Craven Tasker and Moore Uni-Drill direct drills had heavy-duty single disc coulters. The heavy spring-tine coulters on the sixteen- and twenty-row International 511 cultivator drill were considered ideal for use on relatively trash-free heavy land. Another twenty-four-row Tasker grain and fertiliser direct drill with 18 in diameter spring-loaded dished discs to cut a small furrow in unbroken ground had an unladen weight of three tons.

Tramlining kits and drill performance monitors

4.21 The Howard power harrow and narrow spacing Nordsten mechanical feed drill was suitable for medium horsepower tractors.

Electronics solved the problem with an in-cab computer ensuring that the selected coulters received no seed when the next tramline was due. An over-ride switch was used to cancel any additional lifting cycles needed to clear blockages while drilling.

The Agar grain drill monitor provided a means of checking seed flow to the coulters, a task done in earlier days by someone walking behind the drill. Sensor units, one for each set of six force-feed mechanisms had flashing lights and a klaxon to warn the driver of irregular seeding. Agar drill monitor kits were made for drills with up to thirty-six coulters, and sales literature suggested that drilling could continue after dark in the secure knowledge that every seeding mechanism was working properly.

gradually came into use in the mid-1970s. Tramline kits shut off seed flow to selected coulters at pre-set intervals to leave wheelings for subsequent top dressing and spraying operations. In the early days of this technique, and often due to lack of driver concentration, it was not unknown for the odd tramline to appear in the wrong place. This invariably happened in full view of the road!

After a few seasons some farmers who used a direct drill experienced problems with poor drainage and persistent perennial weed growth. Various power harrow/drill combinations, which appeared in the late

4.22 The use of a bridge link was another way to combine seedbed preparation and drilling in a single pass.

1970s, provided the solution by cultivating the soil and drilling the seed in one operation. The late 1970s Lely Combi with a fluted roller feed grain drill mounted above a Roterra power harrow was one of the first machines to apply this new technique.

Amazone, Kuhn and Lely used their own power harrows and drills but most power harrow/drill combinations were a mix and match of different manufacturers' machinery. Such combinations included Dowdeswell, Maschio and Rotomec power harrows working with a Farmhand, Lely, Nodet or Nordsten drill. Most of them were three-point linkage mounted but some, including the Bamlett Western CD and Bettinson CD trailed drills, were hitched by a bridge link to a rear-mounted power harrow.

4.23 Power harrow and pneumatic drill combinations were popular in the mid-1980s.

Trailed and mounted pneumatic grain drills became popular in the mid 1970s. They had a land wheel-driven seed metering mechanism, usually a fluted roller, to supply seed into an airflow created by a pto-driven fan, which carried the seed through a system of tubes to the coulters.

Pneumatic seed drill/cultivator combinations soon followed. Depending on the make and model some were mounted above a tined cultivator arrangement while others were used with a power harrow. Most cultivator/drill combinations were made in

4.24 The mid-1980s pneumatic Overum Tive Drill Jet with a 6 m sowing width had fifty coulters and a 4 ton seed hopper.

4.25 Five models of the mid-1980s Vicon Supaseeder had working widths of between 10 and 26 ft. This 20 ft drill has 58 coulters at 4 in spacings.

continental Europe so, following the introduction of the metric system, drill sowing widths were stated in metres. Most had a 3 or 4 m working width but the widest drills, which were either folded or end-towed for transport, had an 8 m sowing width.

Farmers planning to buy a grain drill in the 1980s had the choice of a plain grain or grain and fertiliser drill for direct, minimal cultivations or conventional drilling. There was also a wide selection of mounted and trailed cultivator/drill combinations with mechanical or pneumatic feed and sowing widths of 8 m or more.

Most drill combinations were rear mounted with the drill mounted above the cultivator or power harrow. The Lely power harrow/drill combination was an exception with a large front-mounted grain hopper to counterbalance the weight of the Roterra power harrow and drill on the hydraulic linkage.

4.26 The front hopper for this Accord pneumatic grain drill counterbalanced the weight of the Lely Roterra power harrow and drill combination.

A fan was used to transfer seed from the front hopper to a much smaller hopper on the drill.

Electronic monitoring systems became standard equipment on pneumatic feed grain drills in the 1980s. Sensors, linked to an in-cab monitor screen, were used to detect grain tube and coulter blockages, interruptions in the airflow, feed mechanism faults and low seed level in the hopper.

Cereal crops were sown in one pass with direct drills and some cultivator/drill combinations but this was not necessarily a practical option on some soil types. Ploughing, cultivating and drilling in a single pass became a reality in the early 1980s when Lely introduced the Condor system with a rear-mounted plough, two Roterra power harrows with packer rollers and a front-mounted drill. The power harrows, one on the left of the tractor and the other on the right and used alternately, were raised and lowered hydraulically with the seed metering chain driven from the land wheel.

4.27 A 120 hp tractor was recommended for the mid-1990s Kverneland Packomat seeder with a four-furrow plough, a 6 ft wide packer roll and the Accord pneumatic grain drill.

Chapter 5

Root Drills and Precision Seeders

Root Drills

Turnip and mangel (or mangold) seed was broadcast by hand until the 1870s when the Rev James Cooke drilled his fodder roots in rows with a modified version of Jethro Tull's drill. James Smyth introduced his cup-feed drill for cereal and roots in 1800 (page 70) and within a few years local blacksmiths were making their own versions of the drill.

Smyth, Hornsby, Garrett, Yates and others made cup-feed drills for root crops from around 1920 after the first sugar beet crop was successfully grown in the UK. Suffolk coulters were used on root drills and each coulter had a rear press wheel to help maintain the even drilling depth required for sugar beet seed.

Mangels and turnips were sometimes grown on the ridge and special coulters were made for this work. Some farmers filled alternate seed cups with putty in an effort to sow the seed further apart in the rows but as the seed was multi-germ two or more plants usually emerged from each seed cluster and hand singling was an important part of growing the crop.

Cup-feed drills for root crops traditionally had four coulters until the late 1940s when James Smyth and other drill manufacturers moved up to five- and six-row drills with a tractor drawbar or steerable fore carriage for horses.

Some smallholders were drilling root crops with a seeder attachment for the Planet push hoe in the 1930s and by the mid-1940s sugar beet was drilled on some farms with four or five Planet seeder units on a tractor toolbar. Imported from America, the Planet seeder unit had a coulter, coverer and rear press wheel and a wavy disc agitator shaft-driven from the front wheel.

The wavy disc agitator metered single seeds through a small round hole in a seed disc located in the bottom of a small seed hopper. Each unit had a set of four seed discs with forty-six different sized holes for drilling any size of seed from clover to peas. The seed disc was

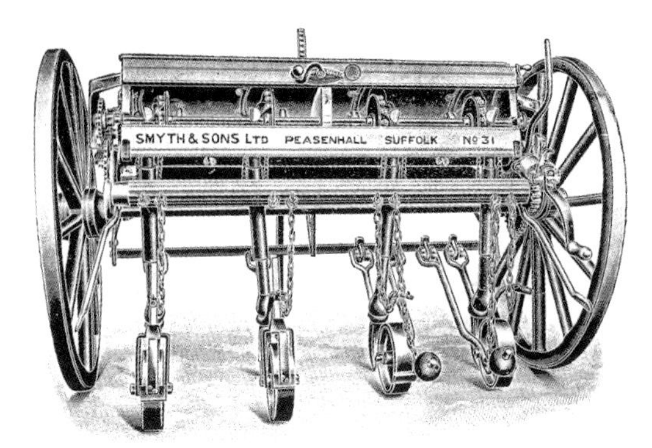

5.1 *Press wheels on this 1950s cup-feed drill firmed the seed in the soil.*

5.2 *Yates root drills could be used on the flat or on the ridge.*

rotated to expose different sizes of hole either to change the seed rate or to drill a different type of seed. Although a degree of seed spacing was achieved with multi-germ sugar beet seed, it was still necessary to single the young plants by hand.

Bean seeder units like the Planet seeder had a land wheel-driven wavy disc agitator in the hopper and a set of four round hole seed discs. Made by Humberside Agricultural Products at Brough in Yorkshire in the late 1940s, Bean seeders were sold as individual units for farmers to use on a suitable three-point linkage toolbar. Humberside Agricultural Products also made a complete three-point linkage drill with a master land-wheel drive through a gearbox to the four, five or six Bean seeder units.

Thomas Green & Son at Leeds made Bean seeder units with the option of a 4 or 11 quart capacity hopper in the late 1950s. They also made the Bean master wheel drive three-point linkage drill and new design of cell wheel seeder unit. The new cell wheel Bean seeders with a 10½ pint hopper and chain drive to the seeding mechanism were used to drill rubbed and graded sugar beet or turnip and swede seed on the flat or ridge.

5.3 Bean seeder units were made for Fordson, Ferguson, David Brown and Ransomes toolbars.

Bean agitator feed and cell wheel seeder units were manufactured by Gloster Equipment at Gloucester in the early 1960s and by RA Lister at Dursley in Gloucestershire in 1963.

Other late 1940s and early 1950s seeder units included those made by MB Wild at Birmingham, the M&G sight-feed seeder and the Catchpole Engineering Taxigraine seeder. Wild seeder units, similar to the Bean and Planet, had a wavy disc agitator to meter seed through a selected hole in one of a set of seed discs. The agitator feed M&G seeder, made by Murwood Agricultural, metered seed through an adjustable hopper outlet to a cast-iron coulter.

5.4 The Gloucester precision seeder unit was demonstrated at the 1963 National Sugar Beet Drilling Trials.

It was standard practice in the days of horse- and early tractor-drawn drills for someone to walk behind to check the seed mechanism. For this reason Murwood advertised the fact that the drill's 7 pint capacity hopper had a Perspex panel to give the person following the drill a clear view of the flow of seed to the coulters.

5.5 The M & G sight-feed drill was used to sow various crops, from onions and parsnips to cabbages and peas.

Precision Seeders

Most early 1950s precision seeders, with the exception of the belt-feed Arden and Stanhay, had a cell wheel mechanism with a single line of equally spaced seed-carrying indentations around the rim. A sales leaflet explained that the Arden seed spacing drill was 'designed by a practising farmer for practical men'. The drill had a pair of shaped belts, which formed a small trough, to carry seed forward from the hopper to the coulter. Sales literature also pointed out that the seed belts were in full view of the tractor driver.

The Webb was the most popular cell wheel precision seeder unit in the 1950s. Others on the market at the time included those made by Milton, Becker, Fähse, Massey Ferguson, Robot, Russell, Twose and Vicon.

5.6 A fertiliser attachment, which placed a band of fertiliser about 2 in to one side of each row of seed, was optional for the early 1950s Robot seed drill made by Transplanters (Robot) at St Albans.

The four-row American Milton cell wheel spacing drill, tested by the Norfolk Agricultural Station in 1949, was marketed in the early 1950s by the Irish Sugar Co as the Armer Milton precision seeder.

Seed was carried from the hopper by a bronze cell wheel with 240 cells in two rows around the periphery. Each cell carried a single seed to an ejector finger where it fell little more than an inch into a shallow furrow. After the seed was covered the rear press wheel firmed the soil. The speed of the cell wheel could be adjusted to sow single seeds between 7/8 and 1¾ in apart and alternative cell wheels were available for different types of seed.

Vicon Monodrill precision seeder units were supplied with a set of large-diameter cell wheels for drilling various crops. It had a high/low ratio gearbox in the drive from the land wheel to the cell wheels and could be used to drill at speeds of up to 5 mph.

Twose seeder units had a revolving plate in the bottom of the seed hopper with a ring of equally spaced holes. A small rotary brush allowed a single seed to fall through

5.7 *Massey Ferguson made four-, five- and six-row versions of the No. 32 cell-wheel precision drill.*

each of the holes in the seed disc and swept surplus seed away to prevent double seeding and blockages.

The introduction of rubbed and graded mainly single-germ seed in the early 1950s prompted the introduction of several new precision spacing drills. The first Stanhay belt-feed precision seeders, introduced by Stanhay at Ashford in Kent in 1953, were sold as individual units with a coulter, coverer and rear press wheel. A small endless belt with a number of equally spaced holes to carry seed from the hopper to the outlet point was shaft-driven from the seeder unit front wheel.

There were two models of Stanhay seeder; the Mk I was used for a variety of crops from onions to sugar beet while Mk II seeders with wider seed belts were used to drill large seeds including peas, beans and maize. Belts with different numbers and sizes of holes for Stanhay seeder units were used for different types of seed and to change seed spacing in the row.

5.8 *Depending on the crop being drilled, the Vicon Monodrill hopper capacity was between 4½ and 7 lb of seed.*

5.9 *The six-speed gearbox on the Webb Space-a-matic master-wheel drive precision seeder provided a choice of six seed spacings in the row.*

About 25,000 Stanhay seeder units had been sold when a pto-driven Stanhay drill was launched at the 1958 Royal Show. It was not a great success and within a couple of years a master wheel drive drill with a land wheel-driven gearbox to vary seed spacing in the row replaced the pto-driven model.

The first Webb cell wheel precision seeders with a set of cell wheels for drilling different sizes of seed and to alter seed spacing in the row were also sold as single units. The Webb precision seeder range in the early 1960s included wheel-driven single seeder units, four- to eight-row pto and master wheel drive drills and special ridge drills. The Webb Space-a-matic master wheel drive drill with the gearbox in the main drive to alter seed spacing appeared in 1964.

Four- and five-row mounted root drills made by Russell's of Kirbymoorside in the late 1950s had a full-width wooden hopper with land wheel-driven rotary seed discs. Four sets of seed discs for different types of seed were supplied with the drill and a five-speed gearbox was used to vary the seed rate.

The mid-1960s range of four- to ten-row Stanhay precision drills had a battery-powered monitoring unit with flashing lights to warn the driver if a belt was not drilling seed. Another light flashed when the seed

5.10 *Introduced in 1960, the Fähse Monocentra cell-wheel precision seeders were later imported by Eric Matthews and then by Catchpole Engineering and Vicon.*

hoppers were almost empty. Optional equipment included conical rollers for drilling on the ridge. The 1971 Stanhay price list included four- to twelve-row master wheel drive drills with MkI and MkII seeder units and a master wheel-drive tandem model with between six and twelve seeder units staggered across front and rear-mounted toolbars.

5.11 An optional fertiliser placement attachment was made for the four- to twelve-row Becker Aeromat precision drill.

Vicon Agricultural Machinery at Ipswich imported the master wheel drive four- to twelve-row Fähse Monocentra cell wheel precision drills from Germany in the mid-1960s. The combination of four different large-diameter cell wheels and a six-speed gearbox gave a range of seed spacings from 1½ to 10 in apart in the row.

5.12 Four-, five-, six- and eight-row Monosem vacuum-type precision seeders were imported by F W McConnel in the late 1970s.

Compressed-air and vacuum-feed precision seeders appeared in the mid-1970s. The German-built Becker Aeromat cell wheel drill used a jet of compressed air from a small pto-driven compressor to hold a single seed in each cell and at the same time gently blow any other seeds back into a holding chamber. On reaching the outlet the air pressure was used to blow single seeds into a prepared shallow furrow. Other drills used vacuum to draw single seeds on to a cell wheel or seed disc.

The Monosem pneumatic precision seeder, marketed by McConnel in the late 1970s, had a rotating seed selector disc which collected seeds from the bottom of the hopper. The vacuum held a single seed against each hole in the selector disc until it reached the seed outlet where the vacuum was released and the seed fell into a shallow furrow.

Stanhay and Webb precision seeders commanded a high percentage of the market after their introduction in the mid-1950s but they did not have a monopoly. David Thomas at Newtown in North Wales introduced a modified design of cell wheel precision seeder in the late 1970s. Seed from the main cell wheel was transferred to a second seed wheel with a peripheral speed equal to the forward speed of the tractor and drill. This allowed the seed to fall to the ground without bouncing which, according the makers, made it possible to drill at 8 mph rather than the more usual speed of 4 mph.

The Belgian-made Tank precision seeder marketed by FA Standen at Ely in the late 1970s with 20 in diameter cell wheels was also claimed to be suitable for high-speed drilling without sacrificing accuracy.

Stanhay and their arch rivals EA Webb of Exning were both part of Hestair Farm Equipment in the mid-1980s. Hestair products at the time included Bettinson direct grain drills, twelve-row master wheel drive Stanhay S981 belt-feed and Webb five-row cell wheel precision drills with optional ceramic coulters.

The Stanhay S870 belt-feed seeder units, with single-, double- and triple-line seed belts and coulters, could be used to drill various types of seed, either in a single row or in two or three closely spaced rows. By the late 1980s, and trading as Stanhay Webb, the range of precision drills made at Exning included the S981 and S870, the multi-row Stanhay Rallye 590 and also the Jumbo precision seeder for drilling any size of seed from sugar beet to maize and ground nuts to broad beans.

5.13 Stanhay's 870 precision seeder units could be used to drill single or closely spaced double or triple rows of seed.

Refinements in the 1990s include the introduction of different designs of rear press wheel to cater for a range of soil types but the most obvious change was the increased number of units on a toolbar. Wider drills with up to eighteen seeder units appeared either mounted on a rigid toolbar or on two centrally hinged toolbars. The widest rigid toolbar drills had an end-tow arrangement while those with a centrally hinged toolbar were folded hydraulically for road transport.

The Stanhay Singulaire for natural, coated and pelleted seed was introduced at the 1990 Smithfield Show. Single seeds are held by vacuum against holes in a rotating disc and on reaching the coulter outlet the vacuum is released from the hole and the seed drops into the soil. A later model of the Singulaire vacuum seeder has multi-line coulters for drilling vegetable crops in a single row or close spaced double and triple rows.

5.14 The Stanhay Singulaire vacuum precision seeder could be used to sow a single row or to drill two or three closely spaced rows.

Chapter 6

Planters

Potato Planters

Potatoes and vegetable plants were planted by hand on most farms in the 1930s and '40s. A horse- or tractor-drawn ridger or plough was used to open furrows for potatoes, the spacing between each tuber usually depending on the size of the planter's boots. After planting, the soil was turned back over the potatoes with a ridger, plough or hand hoes. Splitting back the ridges with a tractor-drawn ridger required considerable concentration on the part of the driver.

6.1 *Ridging with a David Brown Cropmaster. The driver had to keep the wheels on top of the ridges when splitting them back to cover hand-planted potatoes.*

Most vegetable plants were set in hand-dibbled holes or placed in a narrow furrow and firmed with a heavy boot.

One- and two-row mechanical potato planters, mainly from America, saved a great deal of backache when they first appeared in the 1920s. Some carried out the planting operation in one pass using a series of cups on an endless belt to carry single potatoes from the hopper to a discharge point where they fell into a furrow and were covered by ridging bodies. Tuber spacing in the row was controlled by the speed of the cup-carrying belt.

An alternative type of planter with hand-filled cups carried the potatoes to the drop tube where they fell into a prepared furrow. However, Ransomes, which introduced the Ipswich two-row automatic planter in the 1890s, was way ahead of their American rivals. Working on previously ridged land, the two-row Ipswich planter, with two sets of cam-operated fingers, dropped single potatoes at pre-set intervals into prepared furrows.

A German potato planting aid, introduced by Stanhay at Ashford in the late 1930s, was advertised as a trailed dibbling implement. Small shares cut a shallow furrow and star-shaped wheels dibbled deeper holes in the furrow ready for hand labour to drop a potato in each hole.

The Ferguson, Robot, Teagle and other hand-fed planting attachments for mounted ridgers in the mid 1940s made life a little easier at potato planting time. The Teagle potato planter, introduced in 1941, had two endless chains with cups which picked up single potatoes as they moved upwards through the hopper.

6.2 *It was much easier to split back the ridges with a front coverer., the front bobbin wheels kept the ridger bodies on top of the ridges and the tractor wheels ran between the newly formed ridges.*

6.3 This McCormick automatic potato planter was made in America in the 1920s.

The cups tipped the tubers down a spout into a prepared furrow and following ridger bodies covered them. When a cup failed to collect a potato an operator riding on the planter filled the cup to ensure no gaps were left in the row.

The Teagle, which set four to six acres in a day, was claimed to be as accurate as hand planting. Teagle also made two-row dropper type planters for chitted and unchitted seed potatoes in the mid-1940s. According to sales literature children as young as nine could work on the Teagle planter which under average conditions planted between a half and three-quarters of an acre in an hour.

Ferguson and Robot two-row planter attachments for three-row mounted ridgers had two dropper tubes and two seats bolted to the toolbar between the ridging bodies. A bell unit on a small land wheel, which was a guide for the two workers seated on the planter, could be set to ring at intervals of between 8 and 16 in. A coulter at the

6.3A A rack for trays of chitted seed was an optional attachment for the Ferguson ridger and planter.

6.4 The fertiliser attachment on this two-row Ferguson planter for unchitted seed was chain-driven from the tractor wheel.

bottom of the tube made a furrow and every time the bell rang the operators dropped a potato down the tube into the furrow. To complete the job the ridger bodies covered the tubers.

A land wheel-driven fertiliser placement attachment was made for the Ferguson planter which was either used with a hopper for unchitted potatoes or a frame for trays of chitted seed. The Robot dropper-type planter was similar to the Ferguson with a bell to regulate spacing. Potatoes could be planted between 7 and 24 in apart but in the latter case the bell had to be set to ring at 12 in intervals and potatoes were planted at alternate rings of the bell.

6.5 A 1952 advertisement for the three-row semi-automatic Cover-well potato planter.

Mid-1960s potato planters with an automatic tuber spacing mechanism included the Bamlett, Bruff, Cramer, Cover-Well Hassia, Johnson, Massey Ferguson, Packman, the Robot made by Transplanters Ltd at Sandridge in Hertfordshire and the Smallford. Some were hand fed and required an operator for each row while others were semi-automatic planters which collected single potatoes from the hopper and dropped them down a tube to the ground. An operator was needed on some semi-automatic planters to fill any empty cups while others had a small rotating magazine that automatically filled any empty cups.

Two- and three-row semi-automatic potato planters, including the Bamlett, the Robot and the Cover-Well, had a series of hand-filled hinged cups on a land wheel-driven horizontal rotor. The cups carried the potatoes to the outlet point where they were dropped down a spout into the furrow and were covered by a pair of angled discs or ridging bodies.

The three-row Smallford automatic planter first

made in 1953 by Tractor Shafts Co at St Albans needed one person to ride on the machine to operate the controls and fill any empty cups in the three endless conveyors carrying tubers from the hopper to the ground. The central row of cups dropped the potatoes into a furrow made by a coulter at the bottom of the drop tube while those in the left- and right-hand cups were carried sideways by paddle conveyors to the drop tubes.

6.6 *A fertiliser attachment was optional for the mid-1950s two-row Robot planter.*

The Bruff potato planting attachment for various makes of ridger toolbar was made in the mid-1950s at Suckley in Worcestershire. Suitable for planting chitted and unchitted seed, the Bruff planter had two endless land wheel-driven conveyors to carry the potatoes to the dropper tubes where they fell into a prepared furrow. Seats were provided for two workers to place a single potato in each compartment on the belts.

6.7 *Only one operator was required on the three-row Smallford potato planter.*

Transplanters (Robot) Ltd made two-, three- and four-row semi-automatic planting attachments for most makes of mounted ridger. Optional equipment for the Robot planter included a fertiliser placement attachment, a frame for chitted seed trays and a 3 cwt capacity hopper for unsprouted seed. Cooch & Sons at Northampton made trailed and mounted versions of the Robot planter in the early 1960s and it was also made by RA Lister at Dursley in the mid-1960s.

The Johnson two-row hand-fed planter had canvas cups formed between two steel discs to carry the potatoes to the ground. The feed mechanism was chain driven from one of the land wheels and different sized sprockets were used to alter tuber spacing in the row. The two-, three- and four-row planter made by Packman Machinery, and later by Baker Perkins at Twyford in Berkshire, used rotating cups on a horizontal axle to carry the potatoes to the ground.

The drive arrangement kept the cups close together at the top for hand filling and moved them apart as they neared the ground. Spacing in the row depended on the speed of the cups carrying the tubers to the furrows made by the ridging bodies while a pair of angled discs covered the planted crop.

6.8 The Bruff potato planter attachment had two endless conveyors to carry single potatoes to the ground.

6.9 The canvas compartments on the Johnson hand-fed potato planter carried potatoes to the ground.

6.10 *The feed mechanism on the Smallford three-row automatic planter was claimed to plant up to 500 potatoes in one minute.*

Johnson's Engineering at March was still making planters with canvas cups to carry potatoes to the ground in the mid-1960s. Within a few years Johnson's were also selling the German-built Cramer automatic planters and Underhaug Faun hand-fed and automatic planters from Norway.

Cramer planters were added to Ransomes' range of potato machinery when they acquired Johnson's Engineering in 1968. Cramer planters had a separate hopper for each row and cups on an endless vertical chain carried single potatoes from the hopper into a ready-made furrow where they were covered by a pair of angled discs. Any empty cups were automatically filled with potatoes from a small rotating magazine.

The new Ransomes Johnson Chieftain two-row automatic planter, introduced in 1970 and made for five years, had vibrating floors in the hoppers which metered an even flow of potatoes on to vee-shaped endless rubber belts. The belts carried the potatoes to a point above the furrow openers where they dropped into the soil and were covered by angled discs. Tuber spacing was varied with an eight-speed gearbox and an operator rode on the machine to keep an eye on the feed mechanisms.

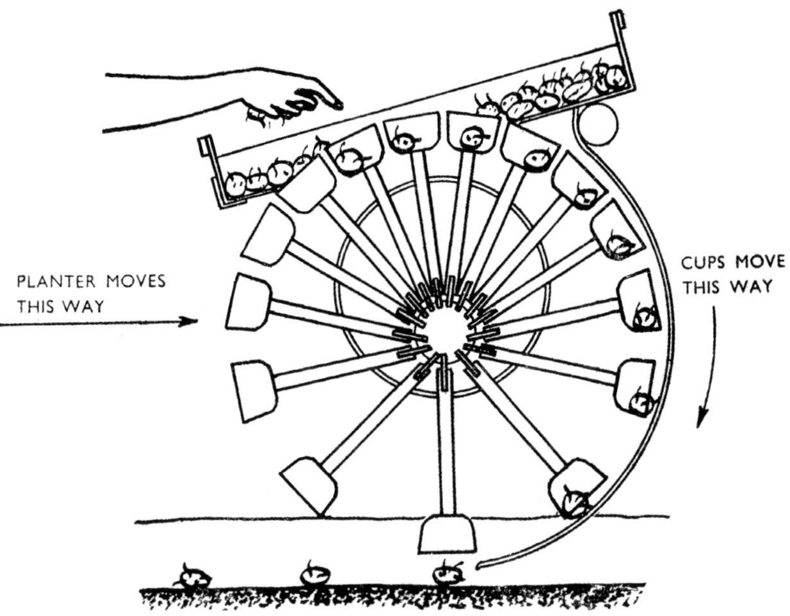

6.11 *The optional fertiliser attachment for the Packman potato planter placed a ring of fertiliser around each tuber.*

Early 1970s potato planters included those made by Hassia, Howard, Massey Ferguson, Mil and Ransomes. Howard Smallford Rotaplanters carried potatoes on angled rubber belts from the hopper to outlet points where they fell a short distance into the ground. An optional fertiliser attachment placed lines of fertiliser on each side of the tubers before ridging bodies covered them. The two- and four-row Rotaplanters had 8 and 16 cwt hoppers respectively and a seat was provided for a worker to check the flow of potatoes from the hopper.

6.12 The Cramer planter had rotating cups which planted potatoes at regular intervals in the ground.

The Smallford Setrite planter, which cost £830 in 1974, could be used with chitted or unchitted seed. Rubber cups on an endless belt carried the seed potatoes from a metering mechanism in the hopper outlet to the ground. Sales literature suggested that two operators riding on the planter could fill any missed cups but more importantly they could, by helping to refill the 3 cwt capacity hoppers, speed up turning at the headland to achieve planting rates of up to an acre an hour.

Hassia planters, the two-, three- and four-row fully automatic planters imported by Melotte Farm & Dairy Equipment, also had an endless chain and cup system to carry potatoes from the hopper to the ground. Different cups were available for various sizes of seed and Hassia planters had the usual rotary magazines to fill any empty cups automatically.

The two-row Massey Ferguson 718 automatic planter, introduced in the late 1950s, had cam-operated, spring-loaded fingers on a large-diameter land wheel-driven disc in both hoppers. The cam mechanism released the potatoes as they reached the outlet chute from where

6.13 John Wallace & Sons at Glasgow imported Cramer potato planters until 1966 and from that date they were marketed by Johnson's Engineering.

6.14 *The Ransomes Johnson Chieftain planter for chitted and unchitted seed worked at speeds of up to 7 mph.*

they dropped into a furrow, to be covered by a pair of angled discs. A fertiliser placement attachment mounted on the tractor and chain-driven from the rear tractor wheel could be used with the MF 718 planter.

Mil Industries made the 718 planter and when it was discontinued by Massey-Ferguson in the early 1970s it was sold as the Mil Autoplanter. It had the same hopper capacity for the potatoes but the fertiliser attachment hopper was front-mounted to overcome the difficulty of fitting the earlier rear-mounted hopper on tractors with a safety cab.

Ransomes imported Cramer and Underhaug-Faun planters in the early 1980s. They lost the Faun franchise when Kverneland

6.15 *The Massey Ferguson 718 automatic potato planter had a work rate of an acre an hour.*

6.16 The two- and four-row Howard Smallford Rotaplanters could be used with chitted or unchitted seed.

6.17 The two-row Smallford Setrite could be used to plant at 26 to 36 in row widths.

acquired Underhaug in 1986 so to fill the gap the firm launched a range of Hassia potato planters at that year's Smithfield Show. When Ransomes closed the farm machinery side of the business in 1987 Cramer potato planters were imported by JF Farm Machines Ltd at Edinburgh.

Virtually all mid-1980s potato planters were unmanned automatic two- and four-row machines with large hoppers and work rate to match. Optional equipment included fertiliser placement attachments, granular insecticide and herbicide applicators and electronic performance monitoring units.

An improved Howard Super Rotaplanter was still made in 1985 when Armer was importing Cramer planters, Hassia planters were being sold in the UK by Pettit and Smallford was making their Setrite and Setronic planters. The two- and four-row Cramer and Hassia planters were fully automatic, a chitted seed version of the Setrite was still available and the Setronic bristled with microprocessors. Sensors, which monitored the belts carrying potatoes from the Setronic hopper, automatically rectified any misses. The potato belts were driven by hydraulic motors and a push button control regulated tuber spacing in the row.

6.18 Ransomes Faun planters in the mid-1980s included two- and four-row machines with half and one ton capacity hoppers. The four-row machine had a work rate of up to 20 acres in a day.

6.19 Melotte imported Hassia planters until 1986 when they were added to the Ransomes range of potato machinery.

6.20 The mid-1990s Kverneland UN3300 six-row trailed potato planter with an optional electronic monitoring unit could plant up to 40 acres in a day.

6.21 Chitted potatoes could be planted in rows or in 2 m wide beds with the Standen Big Boy planter.

Transplanters

Mechanical transplanting machines working at speeds of less than one mile an hour have taken much of the backache from the field-scale growing of vegetables. Several companies, including Accord, Cooch, McConnel, Nicholson, Robot, Russell, Smallford and Teagle, have made either hand-fed or mechanical transplanters at some time since the mid-1940s.

6.22 The Teagle two-row transplanter could be towed from a tractor drawbar or mounted on the three-point linkage.

Some vegetable growers bought a two-row Teagle transplanter complete with a toolbar in the mid-1950s for £80. It had vee-shaped furrow openers and after the plants were placed by hand in shallow furrows the soil was lightly closed by spring-loaded fins and then firmed with a pair of press wheels.

Bundles of plants were carried on trays in front of two operators riding on seats above the press wheels. A third operator seated between the plant trays cut open the bundles, sorted them and handed small bundles of plants to his colleagues. A clicker wheel with eight spacings, similar to that used on dropper-type potato planters, provided a guide to the correct plant spacing.

6.23 The Russell two-row trailed Multi-Planter attachment could be used on a self-lift cultivator.

AM Russell at Edinburgh made two designs of the Multi-Planter from the mid-1940s to the mid-1970s. One was similar to the Teagle planter with a furrow opener and rear press wheels, the other had two hand-fed angled rubber discs, driven by the press wheels, which carried single plants to a shallow furrow. Two, three or four planting units and plant trays were attached to a trailed self-lift or tractor-mounted toolbar.

A team of five people was needed for the three-row Multi-Planter. Two sat on the machine splitting up the bundles of plants and handing them to the other three workers who placed single plants, roots upwards, between the angled rubber discs.

The three-point linkage mounted Rayneplanter, made by Nicholsons of Newark, was claimed to plant up to 4,000 plants an hour. The two-row Rayneplanter came with furrow openers, press wheels and a tray rack for the plants and, as with some other mid-1950s transplanters, each plant was watered from a 23 gallon water tank carried on the machine.

6.24 Six people riding on a four-row Russell Multi-Planter could plant up to an acre in an hour.

6.25 Four people were needed on the Robot transplanter for close space planting at normal tractor speeds.

Trailed and mounted Robot transplanters had an endless chain with pairs of rubber-covered fingers for each row which were opened by cams as they passed across the top of the machine. Operators seated on the planter placed single plants between each pair of fingers, which carried the plants to the ground where the fingers opened and set them into a shallow furrow. A pair of inclined press wheels firmed the soil around the plants. Adjustments were provided to set the plants at any interval between 4 and 64 in apart in the row.

Sales literature explained that, depending on the type of plant and spacings in the row, four operators on a two-row Robot planter could set

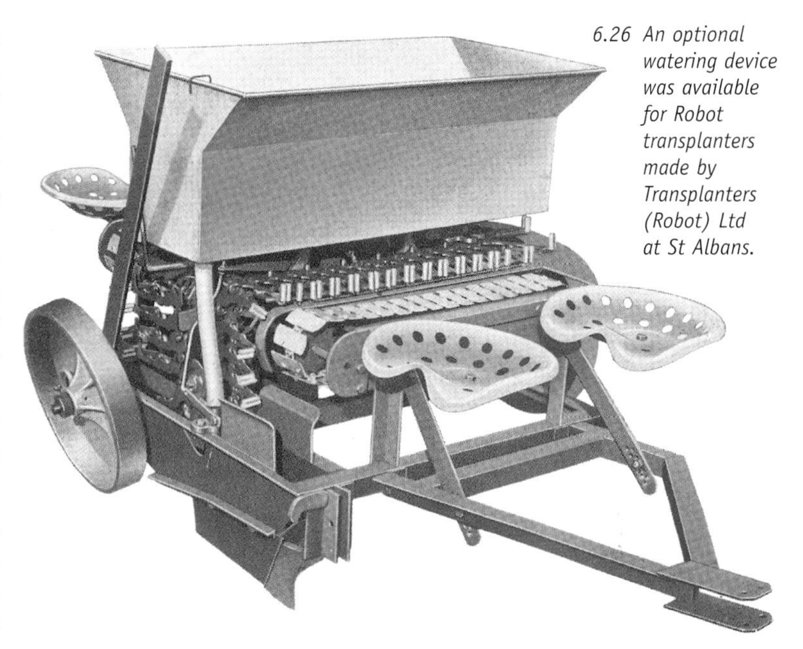

6.26 An optional watering device was available for Robot transplanters made by Transplanters (Robot) Ltd at St Albans.

6.27 With an alternative planting disc the Smallford vegetable planter could be used to plant potatoes. (Arthur Butler)

10,000 plants in an hour. Four people were needed to achieve this output when the plants were closely spaced in the row but two were said to be enough for wider plant spacings.

The Tractor Shafts Company, based at St Albans, made a range of Smallford vegetable, nursery stock and tree planters in the 1960s. The vegetable planter had the usual furrow opener and press wheels. An operator seated on the planter put single plants between four sprung sponge-rubber clamps on the planting disc which carried them to the ground.

A single unit could be used with a small tractor but it was more usual to have two or four units on a tractor toolbar. A set of sprockets was supplied for the chain drive to the planting disc. Depending on the sprocket used and the number of rubber clamps on the disc, the space between each plant could be between 7½ and 48 in.

The technique of planting sugar beet seedlings grown in decomposable pots with a mechanical transplanter attracted considerable attention at the 1981 spring sugar beet demonstration. It was reported at the time that much of the Japanese sugar beet crop was grown in this way but the thought of growing 30,000 plants in separate pots for every acre of sugar beet proved too daunting to contemplate.

Chapter 7

Manure Spreaders

The machinery used to feed and weed farm crops has changed beyond recognition during the last seventy years. Fertiliser was still broadcast by hand on many farms in the 1930s and 1940s and most farmyard manure was spread with a four-tined fork. Farmers used hand hoes and horse hoes to kill weeds. If they happened to own a crop sprayer, it was little more than a wooden barrel on the back of a farm cart with a pump driven by the cart wheel or a farm worker's arm. Chemicals which are now banned, including copper sulphate and dilute sulphuric acid, were among the arsenal of chemicals used to control weeds in arable crops.

A horse-drawn cart or farm trailer and hand forks were the extent of mechanical manure handling on most farms in the 1930s and early 1940s. Manure was carted from yards and stables and dropped off in small heaps across the field to be spread with a fork.

The original Thwaites farmyard manure spreader, introduced in 1945, was an early attempt to mechanise this labour-intensive job. The tractor-towed spreader, with a 7 ft 6 in wide angle-dozer blade on two wheels, levelled small heaps of manure placed at close intervals across the field. The dozer blade spread some of the manure while the rest was swept to one end to be spread by a pto-driven shredder with four swinging chains.

A press report at the time explained that split pins, used to hold the dozer blade in position, were easily removed when putting the blade on top of the machine for transport.

A more efficient Thwaites field heap spreader, later sold as the Wild-Thwaites spreader, appeared in 1947. The trailed and pto-driven field heap spreader had a high-speed spiked drum to shred the manure and an auger with a deflector board to spread it over a narrow strip across the field.

Users were advised that for a dressing of 10 tons an acre the manure should be left in small heaps at 5 yard intervals in rows spaced 5 yards apart. Bigger

7.1 The angled dozer blade on the Thwaites field heap spreader swept manure to a pto-driven shredder with four swinging chains.

7.1A The Wild-Thwaites field heap spreader had an output of 25 to 30 acres in a day.

heaps and two passes with the spreader were needed to apply 20 tons to the acre. A mounted version of the Wild-Thwaites field heap spreader introduced in 1953 was made for the next ten years or so.

Farmers in America were using horse- and tractor-drawn wheel-driven trailer manure spreaders in the 1930s. A few American Massey-Harris and International Harvester trailer spreaders found their way on to British farms in the early 1940s but apart from the Wild-Thwaites field heap spreader there was no other serious challenge to the four-tined manure fork until the end of the decade.

7.1B An early farmyard manure spreader with a tractor drawbar and a horse pole.

Early 1950s wheel-driven manure spreaders included the Bamford FY1, Bentall Galloway, the Hart with an application rate of 2½–16 tons per acre and the Ferguson with a sheet steel body. The McCormick International 200, the Massey-Harris 712 and the Salopian Quickspreader were also on the market.

The Atkinson Spreadall, Salopian Evenspreader and two models of Bloor spreaders made at Crewe

7.2 The Bamford FY3 wheel-driven manure spreader cost £198 in 1958.

were among the small number of pto-driven machines.

The FY1 manure spreader introduced by Bamfords at Uttoxeter in 1949 was one of the first British wheel-driven trailer spreaders. Although most makes of British-made trailer spreader had two shredding drums and a spreading auger, the FY1 and its successors had a third shredding drum below the auger.

Advertised as a fine investment for farmers, the wooden-bodied FY1, which cost £160, spread a 2 ton load in about three minutes. The FY2 replaced the FY1 in the mid-1950s and the FY3, which replaced the FY2 in 1958, was claimed to shred and spread 2½ tons of manure in three minutes. The FY3 spreader had a single control lever and was available with a jaw-type drawbar or a ring hitch. A sluice gate for spreading sloppy manure, complete with a screw adjuster to control its height, was an optional extra.

The 35 cwt Massey-Harris 712 spreader was another early and popular wheel-driven manure spreader. Depending on the speed of the floor conveyor, controlled with a hand lever, the 712 could be set to spread four, eight, twelve or sixteen loads per acre. The wooden spreader body, which held about 35 cwt of wet manure, was gradually tapered outwards at the back to promote a free flow of manure to the spreading mechanism.

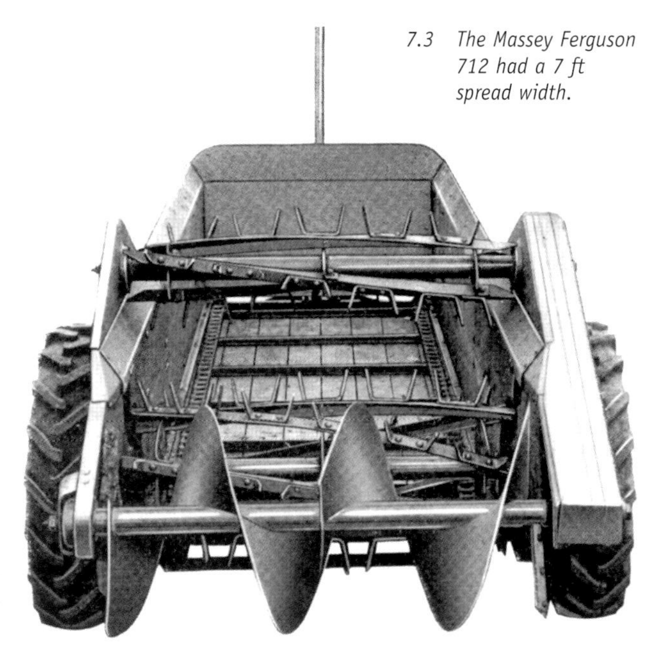

7.3 The Massey Ferguson 712 had a 7 ft spread width.

Following the amalgamation of Massey-Harris and Ferguson in 1953, sales literature for the re-badged MF 712 spreader with a ring and clevis drawbar hitch gave application rates of five, ten, fifteen and twenty loads per acre.

Although the Ransomes & Rapier and James spinning disc spreaders and the Wild-Thwaites field heap spreader were still made in the early 1950s most farmers used a land wheel- or pto-driven trailer spreader. The

7.4 One- and two-ton capacity wheel-driven Quickspreader manure spreaders were made by Salopian Engineers at Prees in Shropshire in the early 1950s.

7.5 *When not required for manure spreading the Bentall Galloway spreader could be converted to a 2 ton trailer in half an hour.*

Ransomes & Rapier manure spreader was similar to a trailed fertiliser broadcaster with a horizontal spinning disc driven by a pair of bevel gears on the spreader axle. It was towed behind a trailer load of manure which was thrown with a hand fork between two guide boards on to the spreading disc.

The James spinning disc spreader, made at Grantham, was also towed behind a cart or trailer. Men riding on the loaded trailer forked manure into a small hopper from where a short elevator dropped it on to a wheel-driven spinning disc.

The American Galloway Trail-r-Spread with a front spreading mechanism was first made under licence in the UK by EH Bentall & Co at Maldon in 1948. The pto-driven Trail-r-Spread had the usual chain-and-slat floor conveyor to carry 2 ton loads of manure forward to twin beaters and spreader fans. The manure was shredded by the hammer mill action of the beaters before it dropped on to two rotary spreading discs.

Sales literature for the Trail-r-Spread explained that front-end spreading had an advantage over rear-end spreading as the manure was thrown down on to the ground over a width of 9–12 ft instead of being tossed up into the air. Another advantage was that as spreading progressed the weight of the manure was always at the front of the machine.

7.6 *The David Brown Albion manure spreader with a six-speed floor conveyor had an application rate of 8–48 tons per acre.*

7.7 A hood attachment for the shredder and a gate for spreading semi-liquid manure were optional extras for the Allis-Chalmers Model 140 spreader.

Over a dozen companies, including Albion, Allis-Chalmers, Bamfords, Bentall, Blanch, Howard, Jones, McCormick International and Salopian, made trailer manure spreaders in the 1950s and 1960s. Most of them were land-wheel driven but a few were driven from the tractor pto.

There were two widths of the wheel-driven Albion spreader; the standard model with a 42 in wide body and six floor conveyor speeds could be set to apply between 7 and 35 tons an acre. The narrower Albion spreader with a 35½ in wide body was used to spread manure in orchards and hop gardens and on potato ridges at a rate of 4–20 loads an acre. Two levers within reach of the tractor seat engaged the drive to the spreading mechanism and the floor conveyor.

Sales literature explained that due to its greater resistance to corrosion the tough and wear-resistant hardwood body of the Albion spreader had a longer working life than similar metal-bodied spreaders.

A later David Brown Albion 2 ton wheel-driven spreader also with a hardwood body, six floor conveyor speeds and a spreading rate of 8–48 tons an acre cost £197.50 in 1959. An attachment for handling 'sloppy manure' was an optional extra.

The Allis-Chalmers Model 140 wheel-driven spreader with application rates of up to 50 tons an acre was made in the late 1950s at Essendine in Lincolnshire. As with other makes, the 2 ton capacity wooden body was a few inches wider at the rear in order to promote the free flow of manure to the shredding and spreading mechanism. Following the amalgamation of Allis-Chalmers and Jones Balers in the early 1960s, the Essendine-built spreader was badged as either the Allis-Chalmers or Jones Model 140 manure spreader.

Bamfords introduced the FS1 spreader in 1960. Intended for the smaller farm it was basically the same as the FY3 but it was narrower and only held 28 cwt of manure.

Announced in 1962, the FY4 with oil-impregnated bronze bearings had fewer greasing points than earlier models and the control lever could be positioned within reach of the driving seat on most makes of tractor.

Dening of Chard, which was acquired by Beyer Peacock in 1951, made Somerset farmyard manure spreaders. The wheel-driven wooden-bodied Dening spreader had a conventional twin beater and auger spreading mechanism. Sales literature explained that five different rates of spread were provided by the control lever used to change the speed of the floor conveyor.

7.8 The McCormick International B 200 manure spreader was made at Doncaster.

The early 1950s McCormick International B200, originally made in America as the 200 manure spreader, had a five-speed floor conveyor which could be set to spread between five and twenty-five 30 cwt loads of manure per acre.

The International Harvester wheel-driven B28-30 and the Swedish-built, power-driven S-125A spreaders joined the long-serving Doncaster-built 30 cwt B200 in the late 1950s. Described in sales literature as 'Big, Low and Robust' the B28-30 had a five-speed floor conveyor with spreading rates of 12½–60 tons an acre. The sales leaflet also pointed out that the wooden floor and sides with steel cappings gave the 2½ ton capacity body maximum resistance to corrosion.

The heavy-duty rotary beaters and spreading auger on the 3 ton S-125A spreader, also with a wooden floor and steel sides, were said to 'tear the manure apart and fling it out in a wide blanket behind the machine'. The B200, B28-30 and S-125A were still included in the International Harvester price list for 1966.

Massey Ferguson inherited the long-serving and reliable wooden-bodied Massey-Harris 712 wheel-driven spreader in 1953. It was still being sold as the MF

7.9 The five-speed floor conveyor on the Dening Somerset spreader could be set to spread 4–20 loads per acre.

712 in 1968, along with the MF 19 power-driven spreader.

The 2 and 3 ton MF 19 spreaders had a one-piece laminated wooden floor and sheet-steel sides. Depending on the forward speed of the tractor the 2 ton spreader, with one spreading rotor and a single-speed floor conveyor, had application rates of 8–48 tons an acre. The 3 ton MF 19, with twin spreading rotors and a two-speed floor conveyor, could spread up to 70 tons an acre in high speed and about 55 tons an acre in low speed.

The New Holland 202 ground drive spreader with steel sides and wooden floor was made by the New Holland Machine Co at Aylesbury in the early 1960s. Sales literature explained that the 'cyclon-action' of the beaters would break up the lumps and spread a light or heavy dressing of manure over a wide area. Optional extras for the New Holland 202 included a fluid end-gate attachment to control the flow of dairy manure and a fine manure attachment with a tray under the beaters to stop poultry manure falling out during transit.

Many farmers in continental Europe bought pto-driven manure spreaders that could also be used as a trailer and this type of spreader became popular on British farms in the late 1950s. Dual-purpose manure spreaders on the market at the time included the Bamford Bombardier, Bentall, Krone Optimat, Tasker Strewer, Vicon Buffalo, Wallace JF and Welger LS60.

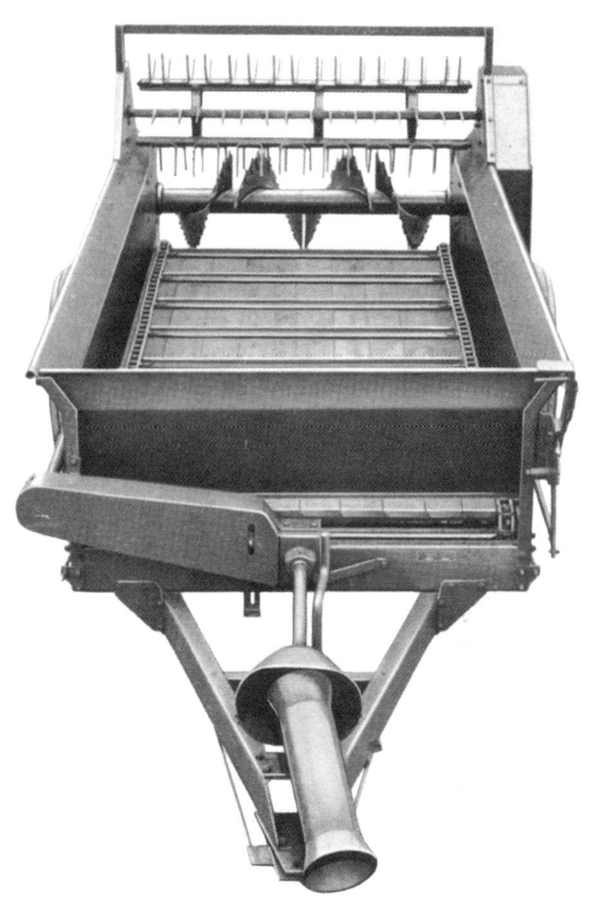

7.10 At a tractor speed of 5 mph the Swedish-built McCormick International S-125A could be set to spread 3–10 loads of manure per acre.

The Vicon Buffalo from Holland was an early example of a trailer manure spreader that could also be used as a self-unloading trailer for root crops or fitted with high sides for carting silage. The rear single-rotor spreading unit on the Wallace UT2 and UT3 trailer spreaders could be lifted off by hand after removing a vee-belt from the machine's five-speed gearbox.

7.11 There was an option of a ring or clevis hitch for the ground-drive New Holland 202 manure spreader.

While there was a longer body on the UT3,

7.12 *There were only six grease nipples on the Massey Ferguson No. 19 power-driven spreader.*

both models held about 3 tons of manure and had spreading rates of 5–30 tons an acre. Attachments included harvest ladders, silage sides, a silage elevator with a cross conveyor, grain sides and a rear unloading beet elevator.

The early 1960s Bamford Bombardier power-driven spreader, which doubled up as a self-emptying trailer, could be used with high sides for silage making. The hinged spreading mechanism was vee-belt driven and could be removed from the trailer in a matter of minutes.

Early machines had a rotary spreading unit but from 1963 there was a choice of a rotary spreader or a flail spreading mechanism for the 3 ton Bombardier. The seven-speed floor conveyor was controlled with a hand lever. The first four speeds were for spreading manure while the others were used to alter the speed of the floor conveyor when using the Bombardier as a self-emptying trailer.

Bentall also made 3 and 4 ton power-driven trailer spreaders in 1960. The all-steel 4 ton contractor's model, advertised as the biggest machine of its type, could be

7.13 *The Vicon Buffalo was also used as a self-unloading trailer.*

7.14 *There was a choice of rotary beaters or flails for the mid-1960s Bamford Bombardier multi-purpose trailer spreader.*

converted from a muck spreader to a 4 ton trailer in less than thirty minutes.

The 3 ton Martin-Markham trailer spreader, new in 1964, was advertised as a genuine three-purpose machine which could be used as a manure spreader with a 7 ft spreading width, a moving floor trailer or as a 3 ton farm trailer. Harvest ladders, side extensions and silage sides were optional extras. The floor conveyor running on a Malayan hardwood floor carried the manure to an easily detached spreading mechanism pto-driven through a five-speed gearbox.

Weeks took a different approach with a manure spreader attachment for their 3½ ton hydraulic tipping trailer which, complete with a parking stand, cost £142 10s in 1965. The spreading rotor on the attachment was pto-driven through a gearbox, and a scoop arrangement pulled by two wire ropes dragged the manure to the spreading rotor.

FW Pettit at Moulton made a similar attachment for their Standard and Victor tipping trailers. The spreading unit, with a five-speed gearbox and stainless-steel belts to convey the manure, could be attached or removed by one man in five minutes.

As with the front-spreading Bentall Galloway Trailer-r-Spread, the Howard Rotaspreader and Taskers Strewer of the early 1960s did not spread manure from the back of the machine.

The Strewer was a side spreader, equally suitable for manure spreading or silage making. A pto-driven flail shredding and spreading mechanism threw manure sideways off the trailer as it was pulled forward along a fixed track on the trailer floor. When empty, the spreader unit had to be returned to the back of the trailer by hand before it could be reloaded.

7.15 *An extended power take-off-to shaft from the tractor was used to drive the manure spreading attachment for the Weeks 3½ ton trailer.*

The Rotaspreader, first made under licence in the UK by Howard Rotavators in 1962, was designed by a farmer in Iowa nd originally made in America by Starline Inc at Illinois. Publicity material for the 4 ton capacity side-delivery Starline Rotaspreader explained that it could be used to spread frozen or hard-packed manure, poultry manure or liquid hog manure in a broad band up to 20 ft wide.

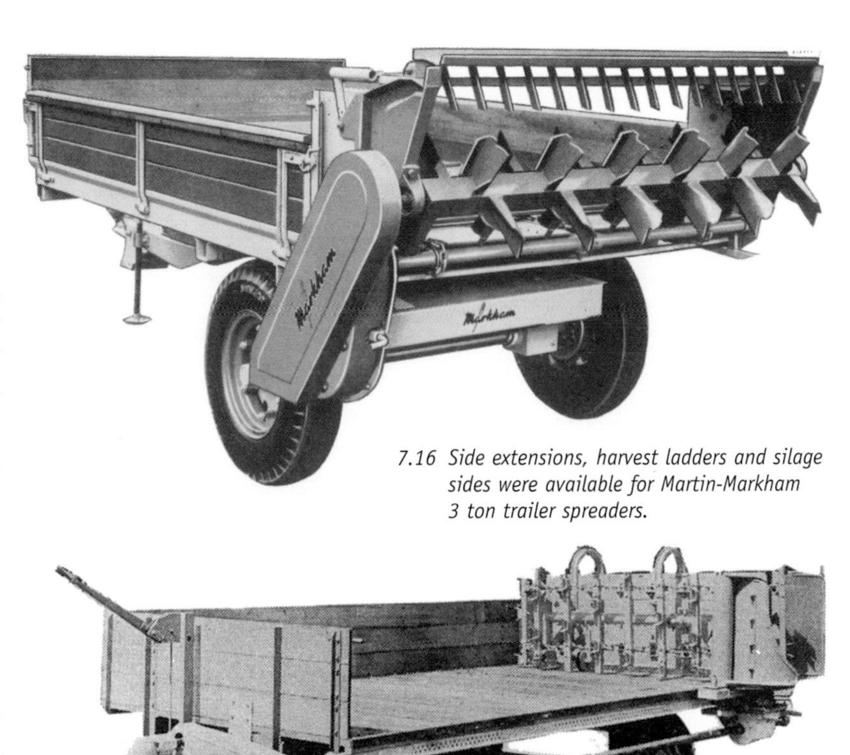

7.16 Side extensions, harvest ladders and silage sides were available for Martin-Markham 3 ton trailer spreaders.

7.17 Early 1960s Taskers sales literature explained that 2,500 Strewers were in use on the continent.

The Howard Rotaspreader had a pto-driven central rotor with flail chains which pulverised the manure and threw it sideways to the ground at a rate of 5–30 tons per acre. The flail chains had to be wrapped around the rotor before loading the spreader with manure, failure to do so usually resulting in a lot of work with a hand fork.

Protected by UK patents, the 3½ and 7 cu yd capacity Howard Rotaspreaders were, with the exception of the John Deere side spreader also made by Howard Rotavators, the only side spreaders on the British market at the time.

Although useful on small farms, dual-purpose trailer spreaders were too small for farmers and contractors with large quantities of manure to spread as quickly as possible. Introduced in 1971, the 7 ton Teagle Titan manure spreader, with a twin rear axle and two swinging flail spreading rotors, was claimed to be Britain's biggest manure spreader.

Most spreaders of the day were pto-driven with shredding drum and auger or flail spreading mechanisms. Early 1970s spreaders included those made by Bamford, Farmhand, Archie Kidd, Massey Ferguson and Vicon.

There was a choice of a toothed spreading auger or flails for the pto-driven 3 ton steel-bodied Bamford MT1 and an optional hydraulically operated slurry door was available for the 3 ton Kidd Fymax spreader. The 6 ton Farmhand 440 and 7½ ton Farmhand 450 manure spreaders could also be used as a forage wagon or self-unloading trailer. The

The 2 and 3 ton capacity No 19 Massey Ferguson spreaders were still in production and Howard Rotavators were making the 3½, 5½ and 7 cu yd capacity Rotaspreader 250.

When the Rotaspreader patents expired in 1977 Howard Rotavators faced stiff competition from several new side spreaders including the Bamford FY 185, the Fernvale Muckchucka, the Kidd Sideflinger, the Sperry New Holland Tank Spreaders and the Weeks side delivery spreader. Howard Rotaspreaders were made at Harleston in Norfolk until Howard Rotavators ceased trading 1985.

Dowdeswell Engineering, which acquired the factory

7.18 The Howard Rotaspreader was introduced in 1962.

7.19 The Kidd 3 ton Fymax spreader was well within the capability of a 35 hp tractor.

7.20 The Teagle Titan, 'Britain's largest muck spreader' with hydraulic brakes cost £760 in 1971.

7.21 A 7½ ton Farmhand manure spreader being filled with a tractor-mounted Twose 190 grab loader.

and manufacturing rights for the Rotaspreader but not the Rotaspreader name, made the renamed Dowdeswell Sidespreader at Harleston. Meanwhile, Farmhand, which had acquired the Howard name, continued Rotaspreader production and introduced new 4.4, 5.5 and 8.6 cu yd capacity Howard Farmhand Rotaspreaders in 1986.

7.22 *Dowdeswell Sidespreaders were made at the old Howard Rotavator factory at Harleston in Norfolk.*

Conventional rear-discharge trailer spreaders were even bigger and more robust in the 1980s when capacities of 7–10 tons were quite common. The 7 ton Kidd Fymax and Farmhand 450 faced competition from several imported machines, including those made by Belarus, Bonhill, Brimont and Zetor. Belarus 5 and 7 ton spreaders with 20 and 24 ft spreading widths from a single auger had work rates of up to 100 tons per hour.

Application rates of 2–30 tons per acre with the 10 ton Bonhill spreader with four vertical beaters were controlled by a variable speed hydraulic drive to the twin floor conveyor.

Front-discharge spreaders, first made by Bentalls in 1960, reappeared in the late 1970s when Colman Agricultural introduced a new range with capacities ranging from 4 to 11 tons. The manure, carried forward by a chain-and-slat conveyor to a large-diameter paddle blade rotor, was shredded and thrown sideways to distances of 30 ft or more on to the ground.

Production of the 6–8 ton capacity range of Colman spreaders was taken over by Richard Western in 1982. Colman spreaders could also be used as self-emptying trailers by reversing the floor conveyor.

Farmers with large amounts of manure to spread in

7.23 *A 75 hp tractor was needed for the 10 ton tandem-axle Bonhill manure spreader.*

the late 1980s used even bigger machines, some with a capacity of 12 tons or more. Four sizes of Umo Belarus spreader had capacities of 5–12 tons, Richard Western made similar-sized rear-discharge machines and the biggest Colman front-discharge spreader held a 10 ton load of manure.

Side spreaders also became bigger in the late 1980s, the three Dowdeswell machines having capacities of 7–12 tons of manure. The larger Fernvale Muckchucka and Kidd Sideflinger both held 9 tons and it took just under 11 tons of well-rotted farmyard manure to fill the biggest Howard Rotaspreader.

7.24 *Up to 100 tones of manure could be spread in an hour with a Belarus 7 ton spreader.*

Slurry Spreaders

Slurry spreaders, or liquid manure distributors as they were known in the 1950s, have a relatively short history. Some Danish farmers collected the run-off from pig and cattle pens in the late 1940s and stored it in underground tanks to be spread with a horse-drawn liquid manure distributor.

7.25 *Colman front-discharge manure spreaders made by Richard Western in the early 1980s had a 30 ft spread pattern.*

A typical distributor consisted of a barrel or tank on a two-wheel cart with a pump either driven by a small petrol engine or chain-driven from one of the cart wheels. The tank contents were pumped into a long trough at the back of the cart from where it was distributed over the ground through a series of holes in the bottom of the trough.

British livestock farmers took little interest in the use of slurry as a free source of plant food until the early 1960s when a few manufacturers, including Alvan Blanch, made slurry tankers. Some had an open top, others had an airtight tank with a vacuum/compressor pump and a rear outlet pipe with a spreader plate or nozzle.

Closed tankers were filled by suction created by the vacuum/compressor pump and the tank was then pressurised to distribute the contents on to a spreader plate or through a nozzle. Open-top tanks were filled from a pit with a pump or a slurry auger and the contents were emptied on to a spreader plate.

A dozen or more slurry tankers, most of them with a 500–700 gallon closed tank, were on the market from the mid-1960s. The Martin-Markham Lincoln Sludgecart, Wright Rain Manurain and the Vicon Hippo were among a number of slurry tankers with a vacuum/compressor pump. The Vicon Hippo with a

Swan Jet nozzle spread slurry in a flat fan pattern over a width of 40 ft. An alternative Angle Jet, used to spread the slurry over fences, lifted it 30 ft in the air spreading it to a distance of approximately 40 ft from the tanker.

Open-top slurry tankers, including the Bamford FY5, the Howard Slurry Puncher and the Salopian Slurribuggy of the early 1970s, contained an enclosed slurry auger or pump. Open-top tankers were emptied by gravity, usually assisted by an auger in the bottom of the tank, either through a nozzle or on to a spreader plate.

Molex and Fulvac tankers were among the many makes of vacuum slurry tanker on the market in the mid-1970s. The Molex had a pto-driven pump with a 30 ft suction lift for filling the 900 or 1,200 gallon fibreglass tank. The same pump was used to empty the contents through a nozzle with spreading widths of up to 60 ft. An optional atomiser jet for the Molex was used to deliver a fine spray of slurry a distance of 150 ft to either side of the machine.

As well as spreading slurry over a width of 40 ft, the mid-1970s Fulvac slurry tanker, made by Allan Fuller at Chepstow, could also be used to agitate the contents of a slurry pit, wash down yards, fight fires or transport bulk liquids.

Slurry tanks without a vacuum/ compressor were usually filled with a pump or a slurry auger but there were exceptions. The Lister Slurry Put bucket attachment for tractor front-end loaders was used to scoop up slurry from a pit and tip it into a waiting tanker.

The Lister Super Scooper was a hydraulically operated self-loading 120 gallon bucket attached to the back of a slurry tanker. The scoop, mounted on a frame similar to a front-end loader, was filled by reversing it into a slurry pit. It was emptied by tipping the slurry into a large funnel-shaped hopper on top of the tank. The Super Scooper slurry tanker, made by Agri-Tech at Whitchurch in the early 1980s, was described in sales literature as the only slurry tanker with a self-loading scoop.

Howard Rotaspreaders equipped with a hinged slurry lid were used for spreading slurry in the mid-1980s. The Slurrymatic loader made for the Howard Rotaspreader by Exmoor Engineering consisted of a 30 gallon bucket on a tubular steel frame bolted to the side of the spreader. The bucket, raised and lowered with an external hydraulic ram, was filled by lowering it into a slurry pit. With the slurry lid open, the contents of the bucket were tipped the into the Rotaspreader.

7.26 The Vicon Hippo Swan Jet had a 40 ft spreading width.

7.27 A high-pressure rotary pump was used to fill and empty the Farrow Ehrwig slurry tanker. It had a 30 ft spreading width and an optional side-spreader nozzle had a maximum jet length of 150 ft.

Chapter 8

Manure Loaders

Farmyard manure was loaded into carts and trailers with hand forks in the early 1940s. However, mechanisation was at hand in the shape of the Dungledozer, the Thwaites hydraulic manure grab and the Painter manure loader.

The early 1940s Dungledozer built around a Fordson tractor with a low-ratio gearbox and Rotaped tracks was by today's standards a contractor's machine. The Dungledozer's rear-mounted rotary shredder was reversed into a heap of manure or stockyard to break up the manure and feed it to an elevator with a swivelling front section. The shredded manure was carried over the tractor and dumped into a waiting cart.

The Dungledozer worked at its best in cattle yards when the manure was at least knee deep. Three tractor trailers or six carts were needed to handle the flow of finely chopped manure when carting it half a mile to the field.

The Thwaites Agricultural Engineering Co at Leamington Spa made the Thwaites manure grab for the Fordson tractor. Introduced to local farmers in 1943, the Thwaites grab consisted of a two-wheeled jib with a grab attached to a wire rope. A hydraulic ram, supplied with oil from a pto-driven pump, opened and closed the grab jaws. The loaded grab was swung round on the jib from the manure heap and emptied into a cart.

The Painter manure loader, designed by a Hampshire farmer in 1944, was made by Kennedy & Kemp at Winchester for the Fordson Model N. It was an early version of the modern front loader with the jib extending a full 10 ft beyond the front of the tractor.

A system of wire cables, controlled by a friction-operated hoist and powered by the tractor belt pulley, was used to raise the loaded manure fork. The tractor was then driven forward to a waiting cart where a tug on a rope tipped the fork full of manure into a cart. Before collecting the next load the jib was lowered to the ground and the tractor reversed to lock the tines in position for the next forkful of manure.

8.1 The Howard Dungledozer could pick up, shred and load up to 15 tons of manure in an hour. (R Merrall)

8.2 The Wild-Thwaites loader winch dragging a forkful of manure to the elevator.

The Ransomes & Rapier loader, introduced in the mid-1940s, was a tractor-mounted crane with a manure grab. When used in an open yard it took less than ten minutes to load up to 2 tons of manure into a waiting cart.

More relief from loading manure with hand forks came in 1946 when Thwaites Agricultural Engineering introduced an engine-powered manure loader. Later sold as the Wild-Thwaites manure loader, it had a hand-steered fork connected by 40 ft of wire rope to an engine-driven winch used to pull large forkfuls of manure to an elevator where it could be loaded into a cart. Two men were needed: one pulled out the fork on its cable, and after it had been pushed into the manure the second man operated the winch clutch while his colleague steered the fork back to the elevator.

The delivery height of the Wild-Thwaites elevator

8.3 The forkful of manure has reached the Wild-Thwaites loader ready to be elevated into a trailer.

was increased to just over 11 ft in the early 1950s. It could be used in livestock yards where there was at least 8 ft 6 in headroom to load manure at a rate of up to 10 tons an hour. Loading could be a one-man operation with the optional remote clutch-control cable attached to the fork handle.

Several different front- and rear-mounted tractor loaders were used on British farms in the mid-1940s. The Ferguson loader with the jib mounted on the front axle was an early example of the type which in 1946, complete with a manure fork and wire rope trip cable, cost £45.

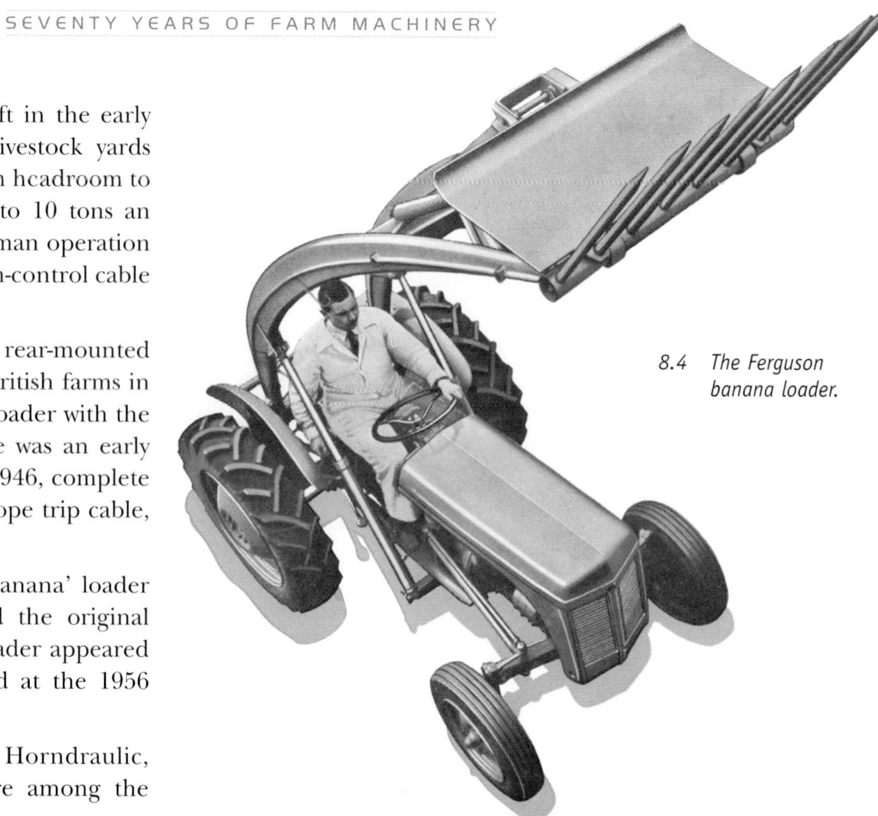

8.4 The Ferguson banana loader.

The Ferguson high-lift or 'banana' loader for the TE 20 soon replaced the original model and the 730 high-lift loader appeared when the FE 35 was launched at the 1956 Smithfield Show.

JC Bamford, Compton, Horndraulic, Markham, Mil and Skyhi were among the

8.5 The JCB Master-Loader for Fordson tractors, made by J C Bamford at Lakeside Works, Rocester, cost £60 in 1952, a manure fork was an extra £15.

companies making front-end loaders in the 1950s and 1960s. The JCB Major-Loader was made between 1949 and 1957 for Fordson E27N Major and Nuffield Universal tractors. The JCB Master-Loader was added in 1951 and the Bamford Si-Draulic loader appeared in 1953.

8.6 *A crane jib was one of the optional attachments for Skyhi loaders.*

The framework of some front-end loaders made it difficult to climb on to the tractor but it was not a problem with the single-arm Si-Draulic which was attached to a sub-frame on the left-hand rear axle housing. Attachments for the JCB Si-Draulic loader included a manure fork, a root bucket, a 10 cwt crane, a bulldozer blade and a hay sweep.

The American-designed Horndraulic loader made by Steel Fabricators at Cardiff and the Skyhi loader made at Isleworth in Middlesex were suitable for most tractors. Attachments for the Horndraulic loader included a buckrake, push-off stacker, crane and an engine-driven hedge cutter.

The 5 cwt manure grab, sack hook and other accessories for the early 1950s tractor-mounted Stanhay hoist jib were raised and lowered with an external hydraulic ram. However, after raising the loaded manure grab, muscle power was needed to swing the jib through an angle of up to 160 degrees before the

8.7 *The JCB Si-Draulic loader, suitable for most popular makes of tractor, could lift 15 cwt to a height of 11 ft.*

contents could be emptied into a trailer or manure spreader.

Most front-end loader booms were attached to brackets bolted to the rear axle of the tractor. The introduction of tractor weather cabs in the late 1950s called for a change of design with new models of front-end loader carried on a frame bolted under the centre of the tractor. Before the days of the diff-lock, wheel spin was often a problem when reversing a tractor with a loaded manure fork in difficult ground conditions.

8.8 The Stanhay hydraulic hoist with a manually swivelled jib was made at Ashford in the early 1950s.

Rear-mounted loaders, including the Cameron Gardner Rearloda and the Protter loader made by AB Blanch at Crudwell, provided an alternative way of loading manure. The Cameron Gardner loader was reversed up to the manure heap to fill the fork and after some necessary manoeuvring the tractor was backed up to a trailer or spreader to dump the load. The hydraulic Protter was also reversed into the

8.9 The Cameron Gardner Rearloda could be left on the tractor while towing a trailer.

8.10 When used with a 60 hp tractor the Parmiter hydraulic rear-end loader could lift 10 cwt of manure to a height of 10 ft.

8.11 The manure fork on the Revol single-arm loader could be swung to either side of the tractor.

8.12 Made in the early 1960s, the Twawye loader could be used in front of or behind a tractor.

heap and, after slewing, loaded the fork sideways through 90 degrees over the spreader. A push-off ram emptied the fork.

In the mid-1970s, rear-mounted manure loaders included the double-arm Cameron Gardner Rearloda and the Bamlett Hijac. Single-arm rear loaders, also mounted on the three-point linkage, included the Knight-Kidd, MIL Mastiff, Parmiter, Revol and Fixol.

The manure fork was raised with an external ram and a trip handle was used to tip the fork into a waiting spreader. The single-arm Revol loader had the added advantage of a swivelling mounting point, allowing the loader arm to be swung through 90 degrees to either side of the tractor before the manure was tipped into a trailer.

Rear-mounted loaders needed plenty of front ballast weight to keep the tractor stable and as a result were never as popular as the front-end loader. Although most rear loaders could be lifted high enough to tow a trailer from the tractor drawbar it had to be taken off before using three-point linkage for other implements.

The Grays of Fetterangus Twawye over-loader offered farmers the best of both worlds. The loader boom could be fitted on its sub-frame to operate either in front of or behind a tractor. There was a full range of attachments for the Twawye loader which sales literature explained could be mounted or removed from a tractor in a matter of minutes.

During the 1970s some farmers with large quantities of manure to handle used a three-point linkage tractor-mounted grab loader similar to diggers used by the construction industry. The more popular models included the McConnel Power Arm with a wide range of attachments and the Webb 520 loader with a 360 degree hydraulic slewing action.

The Twose 190 grab loader, for manure, sugar beet, silage and other materials, had a maximum reach of 15 ft and a height of up to 13 ft 6 in under the grab.

Mid-1970s and early 1980s catalogues published by Massey Ferguson, the Ford Motor Co, David Brown, International Harvester and other

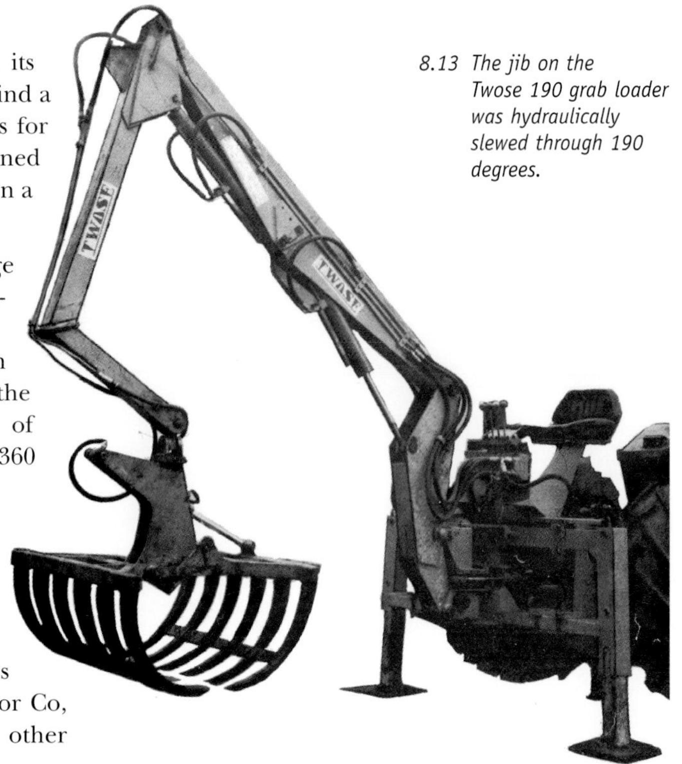

8.13 The jib on the Twose 190 grab loader was hydraulically slewed through 190 degrees.

8.14 The M-F 35 loader for Massey Ferguson 135 and 165 tractors had a lift capacity of 17 cwt to its maximum discharge height of 13 ft 3 in.

8.15 The Farmhand F12 loader on a County 11F tractor with a torque converter had an independent hydraulic system with a pto-driven pump and separate oil reservoir. The F12 had a lift height of 13 ft 4 in and a maximum lift of 1½ tons.

tractor makers listed front-end loaders with higher lift heights and heavier payloads to make full use of increased tractor power.

Other front-end loaders at the time included the Brown Hi-Master, the Cameron Gardner Foreloda, the Farmhand F series, the Quicke, the Mil Marquis and the Steelfab Horndraulic loader. Typical lifting capacities were between 1 and 1½ tons to a maximum height of 11 or 12 ft but the high-lift Farmhand F11 loader and Brown Hi-Master lifted 1½ tons to a maximum height of 17 ft.

The Lawrence Edwards Quicke and Massey Ferguson 80 were described as 'drive-in' loaders that could be attached and detached in a couple of minutes. A parking stand was provided and all the driver had to do was to line up the loader with the brackets on the loader frame, secure the loader boom with two pins and connect the external rams to the tractor hydraulic system.

8.16 The M-F 80 'drive in' loader could be fitted to four-wheel drive Massey Ferguson tractors.

Chapter 9

Fertiliser Distributors

In days gone by farmers bought straight nitrogen, phosphate and potash fertilisers in thick hessian sacks. They were mixed on the barn floor with shovels, the quantity of each constituent in the mix depending on the crop. Sulphate of ammonia, superphosphate and muriate of potash were some of the more common straight fertilisers used in the 1930s and 1940s. Farmers were advised to spread the fertiliser within twenty-four hours of mixing, as it was likely to set hard if stored for longer periods.

Some compound granular fertilisers with various nutrient ratios were available in the mid-1930s but they were more expensive than straight fertilisers and many farmers preferred to save money by mixing their own compounds on the barn floor.

In the mid-1920s considerable quantities of fertiliser were spread from a bucket or similar container usually carried by straps from the worker's shoulders, leaving both hands free to spread fertiliser in a rhythmic pattern. A technical journal at the time noted that there was a great variety of machines available for the distribution of artificial fertilizers. They were more economical than hand broadcasting and were certainly more pleasant for the worker.

Mechanical distributors in the 1930s and 1940s included very basic horizontal spinning disc broadcasters, a full-width spreader with cup-feed mechanisms like those used in a corn drill and the Alexander Jack Imperial artificial manure distributor. The full-width Alexander Jack distributor had a series of scrapers on an endless chain running in the hopper bottom that swept the fertiliser through adjustable size slots on to the ground.

Early types of spinner broadcaster with a small hopper and a land wheel-driven spreading disc were either towed behind, or attached to the back of, a trailer or a horse-drawn cart from where a man shovelled fertiliser into the hopper. The Horwood Bagshaw spinner broadcaster originally made in Australia in the 1930s was manufactured under licence in the UK by Opperman Precision Engineers in the mid-1940s when it cost £47 10s 0d. The price included the driving chain and a sprocket for attaching to the cart or trailer wheel; a range of sprockets was made for use on different sizes of trailer wheels.

Sales literature explained that the machine had

9.1 *The Horwood Bagshaw spinner broadcaster was chain-driven from the cart or trailer wheel. The application of up to 2 cwt per acre was controlled by an adjustable shutter above the spinning disc.*

been made in Australia, and later in the UK, for twenty years and there was not one worn-out Horwood Bagshaw spinner broadcaster to be found in either country.

A late 1940s Ministry of Agriculture and Fishery advisory leaflet listed eight different fertiliser distributor feed mechanisms including endless chain, auger, roller, conveyor and brush, star wheel, plate and flicker and spinning disc. Many of them dated back to the 1930s and both reciprocating plate and plate-and-flicker distributors were still made in the late 1960s. The star-wheel fertiliser feed mechanism was still used with some makes of combine drill in the early 1970s.

9.2 *The Bamford Nu-Drive Supreme and other makes of reciprocating plate fertiliser spreaders were made until the mid 1950s.*

Reciprocating plate full-width distributors, including those made by Bamfords and McCormick International, had three perforated steel plates in the bottom of the hopper.

Depending on the make, either the middle plate or the top and bottom plates were reciprocated by a wheel-driven crank and connecting rod. This action caused fertiliser to fall through the perforations to the ground. A reciprocating agitator above the plates prevented the fertiliser bridging in the hopper and the application rate depended on the length of stroke of the moving plate(s).

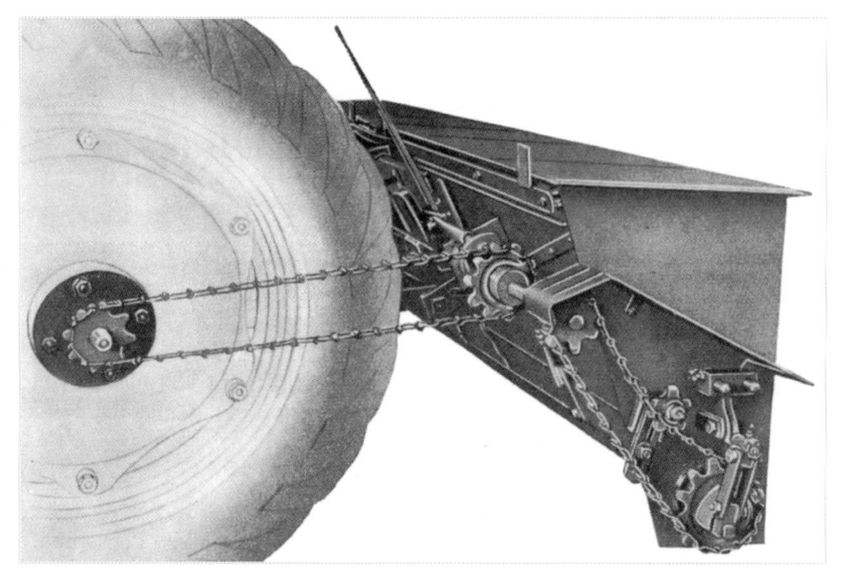

9.3 *Twose roller feed fertiliser distributors applied between 1 and 30 cwt per acre.*

Thorough cleaning at the end of the day was vital as with the combination of fertiliser residues and overnight rain it did not take long for corrosion to wreak havoc in the bottom of the hopper.

The Bamford Nu-Drive distributor had a 'micrometer' application rate adjustment for its 'rustless' reciprocating plates, and optional deflectors were available to spread the fertiliser in wide or narrow bands.

Twose of Tiverton made trailed and mounted roller-feed distributors in the early 1950s. Trailed Twose distributors were land-wheel driven and the mounted models were chain-driven from one of the tractor rear wheels. The roller fed fertiliser through adjustable-sized slots spaced across the full width of the hopper and an agitator bar was used to the fertiliser bridging.

The Knapp New Monarch distributor also had a smooth roller across the full width of the hopper and a rotary agitator. An adjustable scraper board against the roller was used to control the application rate.

Most land wheel-driven plate-and-flicker distributors had seven or eight rotating saucer-shaped discs or plates in the hopper bottom which carried fertiliser to an external row of high-speed rotating fingers where the fertiliser was flicked off the plates on to the ground. Application rate depended on the speed of the plates and adjustable shutters were used to control the flow of fertiliser from the hopper. Plate-and-flicker distributors were easier to keep clean but distribution was erratic on unlevel ground.

9.4 The Massey-Harris 717 plate-and-flicker distributor had forty-eight application rates from 20 to 4,000 lb per acre.

The Massey-Harris 717 plate-and-flicker distributor was made with horse shafts or a tractor drawbar and the most noticeable difference between the Massey-Harris 717 and the M-F 717 was its name. The Lister-Vicon plate-and-flicker distributor with an application rate of 60–3,300 lbs per acre was suitable for use with fertiliser and grass seed. It cost £78 in 1955 when publicity material explained it only took three minutes for a man to strip and clean the machine.

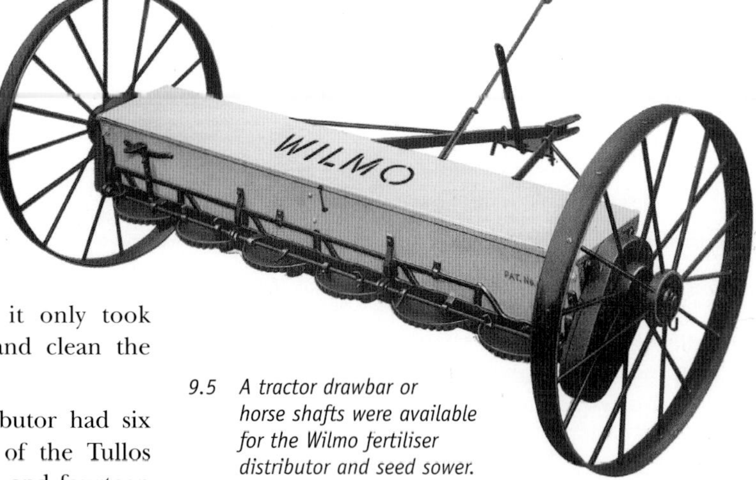

9.5 A tractor drawbar or horse shafts were available for the Wilmo fertiliser distributor and seed sower.

The Wrekin plate-and-flicker distributor had six plates and there were four models of the Tullos Wilmo distributor with six, seven, ten and fourteen plates. The two wider models were mounted on a pneumatic-tyred platform and the plate and flicker mechanism was chain-driven from the wheels. A built-in jacking system was provided to help rotate the hopper through 90 degrees to reduce the overall transport width.

Jack Olding at Hatfield sold Tullos Wilmo distributors in the late 1950s and Alexander Shanks at Arbroath in Scotland and Carlby in Lincolnshire marketed them in the 1960s.

Other types of trailed full-width fertiliser distributor available in the 1950s and 1960s included the Twose Simplex roller feed with a vertical finger bar agitator, the Bisset with a star-wheel feed mechanism and the Warrick fertiliser and seed distributor.

The nine star-wheel spreading mechanism in the bottom of the Bisset distributor hopper could be lowered for cleaning after removing two pins.

9.6 The Tullos Wilmo 14 plate fertiliser distributor cost £266 in 1966.

The 8 and 9 ft wide Warrick fertiliser distributor had an unusual feed mechanism that worked in a similar way to the overhead valve gear on an engine. A set of land wheel-driven rocker arms and push rods opened and closed outlets in the hopper bottom and this action agitated fertiliser down on to a spreader board below the hopper.

Available with horse shafts or a tractor drawbar, the distributor was originally made by Warrick Implements at Chigwell in Essex but by the ate 1950s the Bentall-Warrick distributor was marketed by Bentalls of Maldon. The Warrick distributor mechanism, which could be dismantled for cleaning in under five minutes, had thirty-six settings for small seeds and twenty-four for granular fertiliser.

The spreading disc on the first Teagle trailed spinner broadcasters, launched in 1950, was belt-driven from the land wheels. The application rate was adjusted with a hand lever which raised and lowered a cone above the hopper outlet to control the flow of fertiliser on to the spinning disc. A 1951 Teagle advertisement explained that the broadcaster was built to last, it required no cleaning or housing and could apply many different materials, from 10 lb of grass seed per acre to large dressings of lime, salt and even liquid farmyard manure.

The redesigned Mk2 Teagle trailed broadcaster was introduced in 1953 and the first mounted Teagle broadcasters were added in 1956. There were Lo-Bin

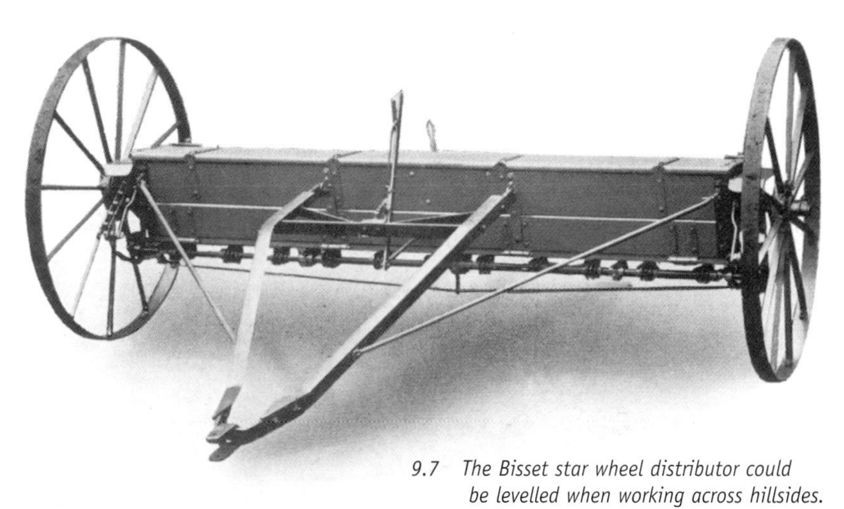

9.7 The Bisset star wheel distributor could be levelled when working across hillsides.

and Hi-Bin versions of the new mounted broadcaster with a belt drive from the pto to a spinning disc but this was changed in 1957 to a direct pto drive arrangement. Although easier to fill, the Lo-Bin model could only be used for granular fertiliser while the Hi-Bin broadcaster with steeper hopper sides was equally suitable for granular and powdered fertilisers.

9.8 The farmer-designed Bentall-Warrick distributor could be used with seed or fertiliser.

Early 1960s Teagle broadcasters included the Versatile with an enclosed bevel gearbox and a removable 8 cwt hopper. The Teagle 1200, with a 12 cwt hopper, claimed to have a spreading width of up to 60 ft and a work rate of 8–10 acres an hour.

The swinging-spout Vicon Varispreader broadcaster with a 5 cwt cone-shaped hopper, made by H Vissers in Holland, was originally sold in the UK by RA Lister at Dursley in the mid-1950s as the Lister-Vicon. Vicon established its own depot at Ipswich in 1958 from where it sold the Varispreader under its own name. The pto-driven Vicon Varispreader, which cost £72 10s, had a removable 6 cwt hopper and a stainless steel swinging spout. A 1 ton trailed Vicon Varispreader was added in the early 1960s.

Lister Blackstone also made their own mounted and wheel-driven spinner broadcasters, the MkII machine with a 3 cwt hopper and a feed shut-off controlled from the tractor seat. It cost £39 in 1958.

9.9 The Teagle Mk.2 trailed broadcaster spread granular fertilisers to a maximum width of 60 ft.

Other spinner broadcasters in the 1950s included the Blanch Triplefit twin-disc broadcaster with a 40–50 ft spread width. The Triplefit could be trailed, attached to the back of a trailer or mounted on the three-point linkage and chain-driven from a rear tractor wheel.

The Grays of Fetterangus Caster was towed either from the tractor drawbar or behind a trailer. The spinning disc was vee-belt driven by a rear castor wheel on the distributor.

The pto-driven Clydebuilt mounted broadcaster and a special belt-driven broadcaster for Massey Ferguson tractors was made by Innes Walker Engineering.

The Lely Crosspreader marketed by Blanch-Lely and later by British Lely had a vertical spreading disc. Unlike other broadcasters of the day, the Crosspreader could be set to spread fertiliser on both or either sides of the machine.

Lely mounted and trailed precision fertiliser broadcasters, pto- or wheel-driven, made by C Van Der Lely in Holland in the early 1960s were marketed by Colchester Tillage. The three Lely broadcasters with a low-level 8 cwt capacity hopper had spreading widths of up to 50 ft in a half-circle spread pattern to the rear or on either side of the machine.

Whitlock Brothers, well known for their red farm trailers, made a spinner broadcaster with a 50 ft spread pattern in the early 1960s. The broadcaster, with a 4 cwt capacity fibre glass hopper and a stainless-steel spinning disc driven by a hydraulic motor could be mounted on the three-point linkage or attached to the back of a trailer.

9.10 *There were mounted and trailed models of the Vicon Varispreader.*

Other early 1960s broadcasters included those sold by Nicholson, Trojan and Lawrence Edwards. The mounted pto-driven and trailed wheel-driven Nicholson broadcasters had a 180 degree spreading arc and achieved a 30 ft spread pattern when travelling at 5 mph.

The Trojan broadcaster, which only cost £29 10s delivered to the farm, had a wooden hopper with no corrosive metal parts.

9.11 *The hopper on the Blanch Lely Crosspreader held 12 cwt of granular fertiliser.*

More than a thousand Spitzenreiter-Automat broadcasters had been sold in West Germany when Laurence Edwards & Co introduced the machine at the 1960 Royal Smithfield Show. The tractor-drawn broadcaster on a two-wheeled chassis had a unique design of spinning disc claimed to ensure complete accuracy under all conditions. It was also said to give an even application without requiring any lumps in the fertiliser to be broken up.

The 24 cwt capacity hopper was sub-divided into four flexible partitions which could be used to carry up to four straight fertilisers, enabling the farmer to mix and spread the required compound fertiliser dressing at the same time. Application rates ranged from 90 to 3,300 lb per acre.

Introduced to UK farmers in 1961, the Swedish-built McCormick International S31-1 fertiliser distributor with a completely new type of distribution mechanism was the forerunner of the modern pneumatic fertiliser distributor.

A pto-driven convex feed plate, under a 9 cwt hopper rotating at 27 rpm, metered fertiliser through six slots into a stream of air created by a fan

9.12 The Lely precision spinner broadcaster could be dismantled for cleaning in a matter of seconds.

running at 1,100 rpm. The fertiliser was delivered to six spouts arranged radially around the fan housing and was broadcast over a width of approximately 20 ft. The application rate was adjusted by using any of the sixty-four different feed plate settings.

Twenty-seven broadcasters and seven full-width distributors were shown at the 1965 Spreaders in Action demonstration in Norfolk. The full-width distributors included the fourteen-disc Wilmo plate-and-flicker machine, the 10 and 12 ft New Idea, the 12 ft John Deere LF marketed by Lundell and the Gandy

9.13 Straight fertilisers in the four-section Spitzenreiter-Automat hopper were mixed and spread in a single operation.

with 6–20 ft working widths imported by Colchester Tillage.

The New Idea had a full-width force-feed agitator mechanism which metered fertiliser through adjustable sized outlets in the bottom of the hopper.

The John Deere distributor used a vertical action 'Propel-R' feed auger mechanism that flicked the fertiliser rearwards.

The Gandy used a full-width auger-type rotor to force the fertiliser through diamond-shaped holes in the bottom of hopper, while adjustable shutters controlled the application rate.

The broadcasters shown at the Norfolk demonstration, with the

9.14 Sales literature explained that four tons of fertiliser could be spread non-stop with a trailer-mounted Whitlock spinner broadcaster.

9.15 The McCormick International S31-1 hopper held up to 9 cwt of granular fertiliser.

exception of the Vicon Varispreader, were single or twin-disc machines with spreading widths from 21 to 50 ft. The Harder Handseat twin-disc mounted broadcaster, imported by Colchester Tillage, had a 10 cwt hopper, application rate was controlled by conical outlets above both discs and, if required, fertiliser could be spread with only one disc.

The 8 cwt Catchpole Twin-Spin with two contra-rotating discs and two propeller agitators in the hopper had a 21 ft spreading width.

The Danish Bogballe spinning disc broadcasters were imported by Ruston's Engineering. They included single-disc mounted models with 6, 8½ and 9 cwt capacity hoppers and a trailed 2 ton Bogballe Model 2000 which could be converted to a self-emptying grain trailer.

The mounted twin-disc Kromag, the single-disc Melodrive and the trailed Hammerbo spinner broadcasters were imported by Melotte Farm & Dairy Equipment in the mid-1960s. Sales literature explained that the Melodrive had an adjustable friction drive from the tractor power shaft to the spinning disc. This enabled the driver to adjust the application rate by varying the speed of the spinning disc without changing gear or the speed of the pto. When spreading in a side wind the hopper could be moved sideways to direct the fertiliser further to the right or left of the disc.

The Hammerbo bulk spreader, with a 2 ton fibre glass hopper, had an endless floor conveyor belt which carried the fertiliser to the twin spinning discs.

The Teagle VM broadcaster, priced at £40, was the cheapest machine at the 1965 Spreaders in Action event but its small 4 cwt hopper needed frequent refilling when broadcasting at its maximum rate of 8 cwt an acre.

Curtis, Padwick & Co at Winchester imported twin-disc German-built Amazone broadcasters in the mid-1960s. The ZA broadcaster with a 30 ft spread pattern had a shear pin in the main drive from the power shaft to the bevel gearbox to prevent stones or pieces of fertiliser bag jamming the discs. A glass eye in the gearbox housing was used to check the oil level and a ZA field test report noted that unlike some other

9.16 The Vicon Multi-Spreader with a 25 ft spread width and a 2 ton hopper was one of the machines demonstrated at the Spreaders in Action event held in Norfolk in 1965.

9.17 The twin-disc Amazone ZA with spreading widths of up to 30 ft and a work rate of 20 acres an hour was also demonstrated at the 1965 Spreaders in Action event.

9.18 A hydraulic motor was used to drive the spinning disc on the Taskers Fertispread. The hydraulic self-loading scoop was an optional extra.

machines on the market the fertiliser did not find its way into the tractor cab.

A demand for more accurate fertiliser application in the early 1970s saw the return of full-width spreaders and the introduction of the first pneumatic distributors. The John Deere, New Idea and Ransomes Nordsten Exact-o-Matic were among the more popular full-width distributors.

The 10, 13 and 19 ft wide Ransomes Nordsten Exact-o-Matic distributor from Denmark had studded rollers which metered fertiliser on to a deflector bar to give an even spread pattern. An optional end-fill hopper arrangement for the wider Exact-o-Matic distributors had a hydraulic motor-driven auger to convey fertiliser tipped from a trailer across the full width of the hopper.

Alpha-Accord, Nodet and Vicon

9.19 The mid-1960s Lister Fantail broadcaster was claimed to spread up to 16 cwt per acre at 5½ mph.

9.20 The 33 ft wide spreader boom on the Accord pneumatic distributor was folded forward for transport.

were some of the more popular mid-1970s pneumatic distributors. Fertiliser was mechanically metered from adjustable hopper outlets into an air stream created by a pto-driven fan. The air stream carried the fertiliser through a system of pipes to outlets with spreader plates spaced across the width of the folding boom. Most pneumatic spreaders had 30–40 ft spreading widths and application rates of up to 12 cwt per acre.

Extreme accuracy was claimed for the mid-1970s mounted and pto-driven Lister Spread twin-hopper precision fertiliser spreader. Augers carried the fertiliser along two full-width booms to forty outlets spaced across the 20 ft spreading width and any surplus fertiliser was returned to the hopper. The application rate could be varied between 1 and 8 cwt per acre when driving at the recommended working speed of 5½ mph.

9.21 The hopper on the mid-1980s Lely broadcaster held a ton of fertiliser.

9.23 The mid-1970s Vicon Aerospreader had a butterfly valve to control the suction used to draw fertiliser into the air stream which carried it to the distributor head.

Tractor loaders and forklift trucks became an integral part of fertiliser handling in the early 1980s as 10 and 20 cwt big bags gradually replaced 1 cwt sacks. Spinning disc broadcasters and pneumatic distributors had bigger hoppers to accommodate the contents of big bags and spreading widths were increased to match.

The mid-1980s Lely single-disc mounted broadcaster had a 36 ft spreading width and, depending on the type of fertiliser, the hopper capacity ranged from 7 to 20 cwt.

Other spinning disc broadcasters at the time included six models of the mounted Vicon Varispreader with 20–40 ft spreading widths and polyester hoppers with a capacity of 8–20 cwt of granular fertiliser. At the other end of the scale four models of the trailed Bamlett Bredal lime and fertiliser spreader had a land wheel-driven floor conveyor to carry fertiliser to the twin spreading discs which, depending on material, also had maximum spreading widths of 20–40 ft.

Accord, Bamlett-Tive, Nodet and Ransomes Nordsten pneumatic distributors were on the market in the mid-1980s. The mounted New Accord distributor with a 1,000 rpm pto-driven fan and the

9.24 Fully enclosed augers at the outer ends of the booms on the Lister Spread distributor carried surplus fertiliser back to the hoppers.

9.25 The 40 ft boom on both mounted and trailed Ransomes Nordsten Air-o-matic fertiliser distributors folded down to under 8 ft for transport.

9.26 One disc on the twin-disc Vicon Rotaflow could be shut off when fertiliser was being spread near a ditch.

9.27 The Overum Tive Jet spreader booms were folded hydraulically for transport.

trailed Ransomes Nordsten 40 ft pneumatic distributor had application rates of up to 12 cwt per acre. Five models of the Bamlett-Tive distributor with a pto-driven fan and wheel-driven metering mechanism had similar application rates and spreading widths.

In the late 1980s some farmers spread their fertiliser with a pneumatic distributor while others used one of the latest and more accurate spinning disc broadcasters. This new generation of single- and twin-spinning disc broadcasters had larger capacity hoppers, improved distribution patterns across an adjustable 30–75 ft spreading width, hopper weigh cells and one-sided spreading when working close to a ditch or field boundary.

Computerised monitoring and control systems were coming into use on broadcasters and pneumatic fertiliser distributors in the late 1980s. Application rates, changed with the touch of a button, could be continuously monitored with the results displayed on a screen in the tractor cab. With ever more sophisticated computer systems, the driver can now enter spreading width, application rate and required forward speed into the computer memory.

A radar sensor measures the tractor's forward speed and an electronic control automatically opens and closes the feed gate slightly to compensate for any variation in the application rate.

Chapter 10

Crop Sprayers

The application of selective chemical weedkillers can be traced back to the late 1890s when copper sulphate solutions were used to kill some of the weeds growing in cereal crops. DNOC, which was found to control caterpillars on fruit trees, also dates back to the late 1800s and in 1911 it was discovered that dilute sulphuric acid killed certain cereal crop weeds.

By the early 1930s some farmers were using a simple crop sprayer to apply these now banned chemicals to kill charlock and a few other cereal crop weeds. They were also controlling potato blight with Bordeaux mixture.

Trials with diluted sulphuric acid applied at a rate of 100–150 gallons an acre on cereals showed that in good conditions it was possible to achieve an almost complete control of weeds such as charlock, chickweed, speedwell and cleavers. The trial report also noted that a sprayer used for these chemicals required an acid-resistant lead-coated copper tank and copper pipes. The pump on horse- or tractor-drawn sprayers was either hand-operated by a man riding on the sprayer or chain-driven from one of the wheels.

Pierce Victor sprayers made by Philip Pierce & Co at Wexford in the late 1940s with either horse shafts or a tractor drawbar had 4 ft diameter steel wheels and an 80 gallon oak barrel. The Victor sprayer was used with pendant nozzles for controlling potato blight or a 20 ft wide boom for spraying acid to burn off potato haulm or kill charlock in cereals.

The wheel-driven brass and phosphor bronze pump was put into and out of gear with a foot pedal and a hand lever used to tilt the boom also controlled the pump bypass valve to start or stop spraying. Pierce Victor horse-drawn sprayers were still made in the early 1950s when an acid spraying attachment was available as an optional extra.

The 50 gallon Massey-Harris No 11 Traction sprayer for one or two horses, made in America in the early 1930s, had a land wheel-driven pump and agitator to keep the tank contents well mixed. Mainly used in rowcrops at 25–36 in spacings, the working instructions for the No 11 sprayer explained that its twelve height-adjustable drop leg nozzles could be used to spray four rows at a time with one, two or three nozzles over each row.

The Four Oaks Spraying Machine Co at Sutton Coldfield in Warwickshire, one of the pioneers of ground crop spraying having won silver medals for their machines in 1912, introduced the horse-drawn

10.1 The Pierce Victor sprayer was claimed to give the best possible insurance against potato blight.

Farmfield charlock and potato sprayer kit in 1941. It consisted of a brass hand pump and the brass boom to be attached to the back of a farm cart but it was left to the farmer to provide a barrel or cask for the spray chemical.

An advertisement for the Farmfield sprayer depicted the horseman operating the sprayer pump while riding on the cart with his back to the horse which was apparently left to make its own way across the field. The standard Four Oaks Farmfield sprayer kit without a wooden cask cost £18 2s 6d. A special acid-resistant alloy tank for spraying a 7 per cent solution of sulphuric acid to control weeds in cereal crops was an extra £4 16s 6d.

10.2 Four Oaks advertised the New Farmfield sprayer in 1941.

W Weeks at Maidstone also made one- and two-horse sprayers in the early 1940s. The Weeks sprayer for insecticides and fungicides had a wooden barrel mounted on a wheeled chassis and the pump was driven from a gear wheel on the sprayer axle.

Barclay, Ross & Hutchinson at Aberdeen also made horse-drawn sprayers in the late 1930s. They had a 100 gallon oak cask, a wheel-driven pump and twenty nozzles spaced across a 16 ft 6 in boom which could be set at a height of between 12 and 42 in above the ground. The steel sprayer wheels could be set at a maximum track width of 5 ft 3 in for applying sulphuric acid, copper sulphate or Bordeaux mixture to rowcrops.

Tractor-drawn sprayers on pneumatic tyres were made by Barclay, Ross & Hutchinson in the early 1940s. Sales literature described them as compressor-type sulphuric acid spraying machines with a compressor chain-driven from one of the wheels to provide the necessary working pressure in the copper tank. A hand-operated diaphragm pump or optional pto pump was used to fill the sprayer tank. The boom and nozzles were similar to those on the earlier horse-drawn sprayers and a seat was provided for the sprayer operator, leaving the tractor driver to concentrate on his driving skills.

Tractor-drawn crop sprayers with hydraulic nozzles, atomisers and air-pressure systems were coming into use in the mid-1940s. Air-pressure sprayers, which worked by pressurising the contents of an airtight tank, applied between 60 and 120 gallons per acre. However, most tractor-mounted low-volume sprayers were little more than a steel tank with a filter basket, a pto-driven gear pump, an on/off valve and a steel spray boom with brass or porcelain nozzles.

The Ransomes Agro Sprayer atomiser sprayer designed in conjunction with the ICI Engineering Research Department at Billingham was made by Ransomes at Ipswich under licence from Plant Protection Ltd. The mounted, close-coupled and trailed models of the Ransomes Agro Sprayer were used to apply a fine chemical mist at a rate of 10–15 gallons per acre.

The Agro Sprayer with a 5 psi working pressure was not only one of the first British-made low-volume sprayers but also a pioneer of the air-assisted spraying

10.3 Twin 40 gallon saddle tanks supplied dilute chemical to the 20 ft spray bar on the nine-row Ransomes Agro Sprayer, which folded down to 7 ft for transport.

technique. Small quantities of atomised chemical, at ten times the normal spraying strength, were introduced into an air stream created by a large pto-driven fan and carried in the air stream to a row of pendant nozzles suspended from a tubular boom.

Most farmers were spraying their crops with a tractor-drawn, low-volume fan jet sprayer in the early 1950s. Others employed a specialist contractor with a large tractor-drawn sprayer, an aeroplane or a helicopter to apply insecticides and fungicides to cereals, beans, potatoes and some other farm crops.

A contractor-owned Auster Autocrat plane fully equipped for spraying cost around £1,500 and, when flying at 60–70 mph at a height of 2–6 ft above the ground, it sprayed about 30 acres an hour. Helicopters were more expensive to run but in favourable conditions could spray 500 acres or more in a single day.

Many companies, including Allman, Dorman Simplex, Drake & Fletcher, Evers & Wall, Four Oaks, Fisons Pest Control, Hart, Ransomes, Romac, Vigzol and Weeks, made crop sprayers in the late 1940s and early 1950s.

EJ Allman, who had started a car repair business at Birdham near Chichester in 1919, introduced the Speedesi tractor-mounted crop duster in 1945, and the Allman Plantector, the first British-built tractor-mounted low-volume hydraulic nozzle sprayer, appeared in 1947. By the early 1950s the range of Allman sprayers included the high/low volume Plantector for the Ferguson TE 20 and other light tractors and the Genimec Spray-Rig kit for use with an ordinary 40 gallon galvanised steel barrel.

The kit consisted of a pump with automatic tank agitation and nozzle cleaning system, a spray bar, a non-drip control valve and taps to shut off individual spray boom sections. The Genimec rig cost £49 10s when it was launched at the 1951 Royal Show but with a high demand for the kit the price was reduced to £44 10s for the 1951 Smithfield Show.

The Allman 60 and the Sprayall 50 sprayers were launched in 1955. The medium/low volume Sprayall 50 had a 50 gallon brass tank, a three-part copper spray boom folded with a rope from the tractor seat and a roller vane pump mounted on the pto shaft. The Sprayall 50 was used to control weeds and burn off potato tops with dilute acid.

10.4 *An optional 20 ft suction hose was available for the Allman Genimec Spray-Rig.*

10.5 *The mid 1950s Allman Speedispray 40 had application rates of up to 40 gallons per acre when spraying at 4 mph.*

10.6 *The pump on early 1950s Dorman low-volume sprayers was driven by a 1 hp JAP petrol engine.*

The mounted Allman 60 with the Genimec control system, four-stage filtration and a 60 gallon galvanised circular tank had a two-part folding spray boom.

The Universal 100, introduced in 1956, was offered as a mounted or trailed 100 gallon sprayer with a 20 or 30 ft spray bar. Other new Allman sprayers in 1956 included the medium/low volume 40 gallon Genimec and the 40 or 60 gallon Genimec 60 sprayers with galvanised barrel-shaped tanks, a roller vane pump and an 18 ft boom.

The 1956 Smithfield Show was the launchpad for the Allman Speedispray 40, also with a roller vane pump. Although designed with three wide-angle jets giving an 18 ft spraying width, it was also used with twelve fan jets on a 19 ft 6 in boom. An advertisement suggested that at last there was 'no need to borrow or hire a sprayer as all farmers could afford to own this sensational new low-priced sprayer'.

Dorman Simplex Sprayers at Cambridge entered the sprayer market in 1948 with horse- and tractor-drawn 100 gallon trailed sprayers and a tractor-drawn

10.7 *The Dorman Sprayer Co made sprayers with a front-mounted spray bar in the early 1950s.*

10.8 The 60 gallon Dorman Utility low-volume sprayer with a galvanised steel tank, and an 18 ft plastic spray boom with non-drip nozzles cost £77 in 1960.

200 gallon model. Both had a small Villiers petrol engine to drive the 12 gallon per minute pump mounted in front of the cylindrical tank and a paddle agitator to stir the contents of the tank.

A 250 gallon high/low volume trailed Dorman sprayer with an adjustable rowcrop axle was added in 1951. A year later the Dorman sprayer range included 50–500 gallon high/low volume trailed sprayers, two low volume trailed models for spraying acid, a sprayer with a front-mounted spray bar and a 50 gallon low volume sprayer for the Land Rover.

The smallest 50 gallon trailer Dorman sprayer came with a welded steel tank and a 21 ft spray bar with anti-drip nozzles. It had application rates of 5, 10 and 15 gallons per acre at 5 mph, which the instruction book explained would be twice as much when driving at 2½ mph.

The early 1950s Dorman sprayer, with a front-mounted boom with flat fan nozzles and a 50 gallon tank on the hydraulic linkage, was made for Ferguson, Fordson and Nuffield tractors. It was pointed out in sales literature that the front-mounted spray boom had the advantage of providing the driver with full visual control of the spraying operation.

The trailed Watson Farmer crop sprayer for low, medium and high volume spraying of insecticides, weedkillers and acids was made by Drake & Fletcher at Maidstone in Kent in the late 1940s. It had two hardwood barrels mounted side by side on a high-wheeled chassis. A 1¾ hp petrol engine was used to drive the pump and the adjustable 16 ft or optional 30 ft boom could be set to its maximum spraying height of 4 ft 8 in.

The early 1950s Drake & Fletcher trailed Minicrop sprayer on a pair of small pneumatic-tyred wheels was made for the smaller farm. An advertisement for the Minicrop with a 60 gallon tank, a 20 ft spray bar and application rates of 5–100 gallons an acre pointed out that when spraying at high volume the driver would need to spend a great deal of his time re-filling the tank.

Drake & Fletcher made the General and Mystifier range of crop sprayers in the mid-1950s. The low volume General sprayers with a 60 or 80 gallon plastic tank and no-drift nozzles cost £82 and £95 respectively. The 50 and 100 gallon Drake & Fletcher Mystifier crop sprayer were high/low volume machines with a 20 or 24 ft spray bar.

The Hart Agricultural Engineering Co at Birmingham made a range of low and medium volume mounted sprayers in the late 1940s. They included the 40 gallon Junior sprayer for Ferguson

10.9 The range of crop sprayers marketed by Boots Farm Sales in the early 1960s included the Hectacare 1600 with a 350 gallon tank and a 33 or 39 ft wide spray boom.

tractors and the 60 gallon Senior for the New Fordson Major and similar tractors. Both sprayers had a gun metal gear pump, copper spray booms and ceramic tip fan jets.

A mounted universal crop sprayer, made by the Agricultural Division of Romac Industries in the late 1940s, had a 50 gallon tank pressurised by a tractor-mounted Romac compressor. Diluted chemicals were applied at a pressure of up to 125 psi from twelve nozzles on the 18 ft wide spray boom. The high-pressure tank could also be used as an air receiver for inflating tyres, paint spraying and other workshop duties when not needed for spraying crops. An advertisement explained that, with whitewash or lime wash in the tank, the Romac sprayer could be used to spray farm buildings and cowsheds.

The 100 gallon trailed Bristol sprayer, made by the Agricultural Division of Welding Industries Ltd at Bristol, had a gear pump and twenty cone jets spaced across two 10 ft galvanised steel booms. The tank held enough dilute chemical to spray 14 acres before refilling, and the pump was driven by the pto or a 1 hp JAP engine.

Arco sprayers, made by Hardy & Collins at Boston in Lincolnshire in the late 1940s, included 150 and 250 gallon high/medium or low volume trailed machines with a pto-driven gear pump, 30 ft booms and ceramic cone nozzles. The Arco 40M low or medium volume mounted sprayer with a 40 gallon steel tank, gear pump and 18 ft boom was priced at £58, a self-filling attachment adding £24 to the bill. The high/medium or low volume Arco 100 with a 25 ft boom, gear pump, built-in self-fill system and application rates of 15–80 gallons an acre cost £135.

A section in the Arco catalogue concerning sprayer maintenance pointed out that when putting the sprayer away for the winter all the pipes should be drained and particular attention should be paid to draining the pump.

Although mainly concerned with horticultural spraying equipment, Four Oaks made a range of eight mounted and trailed crop sprayers for most popular makes of tractor in the 1950s. They included the mounted or trailed Shamrock 333, 444 and 555 sprayers with 40 or 60 gallon barrel-shaped galvanised tanks and a 16 ft spray boom.

For larger acreages, farmers could buy the Four Oaks Major high/low volume trailed machine with 40 ft boom and a 200 or 300 gallon galvanised tank. The mounted Shamrock 333 cost £76 10s in 1960 when the 300 gallon Major was £291 10s.

Leading dairy equipment manufacturer Gascoignes at Reading added sprayers to their product range in 1955. Built to an Evers & Wall design, the Gascoigne sprayer range included 40 and 65 gallon low volume sprayers and a 100 gallon medium/high volume model which cost £65, £107 and £137 10s respectively. Spraying pressure was adjusted from the tractor seat and the spray booms with anti-drip nozzles had a safety break-back mechanism.

A major sprayer manufacturer, Evers & Wall at Lambourn near Newbury, included a sprayer kit for the Land Rover in their 1950 range. A 65 or 100 gallon tank was supplied but, if they wished, customers could use their own 40 gallon barrel. The pump, mounted between the front seats, was driven by the Land Rover's central power shaft and the folding 19 ft spray boom was fixed to the front bumper.

The Sprayquik mounted sprayer, with a brass tank and fittings, was made by Patrick & Wilkinson at Belfast in the late 1940s. The Sprayquik was advertised as the most efficient and economical machine for spraying potatoes, corn and grassland. It

10.10 Optional equipment for Four Oaks Shamrock sprayers included pendant legs for spraying potatoes.

10.11 The Fison's Universal Major had an engine driven gear pump.

10.12 Fisons Pest Control made Ferguson and David Brown low volume sprayers.

It was claimed to be one which could be used by inexperienced workers with perfect ease and safety.

Fisons ran a major spray contracting business, Fisons Pest Control at Cambridge. In the late 1940s and early 1950s they also made Weedmaster and Super Weedmaster low volume mounted sprayers with a gear pump direct-mounted on the pto. The 30 gallon Weedmaster had a 16 ft 6 in boom and the 60 gallon high/low volume Super Weedmaster had a 25 ft spraying width. Fisons Pest Control also made the 200 gallon Universal Major trailed sprayer with an engine-driven gear pump and a 25 ft boom.

In the 1950s Fisons Pest Control were also making the David Brown, Ferguson and Massey Ferguson mounted crop sprayers with a 19 ft 6 in spraying width. The 45 gallon Ferguson sprayer and 50 gallon David Brown had levers on both sides of the tank to fold the hinged sections of the spray booms for transport.

The Ransomes MkI Cropguard trailed sprayer, which superseded the Agro Sprayer in 1949, was in turn replaced by a new range of trailed and mounted Cropguard Junior, Standard and Senior sprayers in 1955. They had 30, 50 and 100 gallon galvanised tanks respectively and, depending on the model, had a gear or roller vane pump mounted on the pto shaft. Flat fan jets were used throughout the range and the 17–32 ft wide angle-iron spray booms were protected by a spring-loaded break-back mechanism.

The Vigzol Oil Co made the Blitzweed low volume sprayer and the Yeoman range of 40 and 60 gallon mounted sprayers in the late 1950s. The specification included a gear pump mounted on the pto shaft, three-stage filtration and anti-drip nozzles. The boom height was adjusted with cam locks used to hold the spray boom on the frame. The earlier Vigzol 40 and 60 gallon Yeoman crop sprayers were sold under the Amoco Vigzol brand name as the Yeoman MkI and MkVII in the late 1960s.

The change from steel to plastic materials in the mid-1960s extended the working life of farm crop sprayers. Dripping nozzles and spray drift had posed problems and the growing use of anti-drip devices and low-pressure non-drift nozzles were welcome improvements in sprayer design. The manufacturers also recognised the importance of spray booms that did not bounce and yaw on unlevel ground. Better designs in the mid-1960s also helped to solve this problem. The safer environment provided by tractor weather cabs, which protected the driver from harmful exposure to spray chemicals, was an added bonus.

There were fewer makes of crop sprayer on the market in the mid-1960s when the leading manufacturers included Allman, Croptex, Dorman, Evers & Wall, Ransomes and Simplex. The eight-

10.13. Farmers were able to choose a medium/low or high/low output Ransomes Cropguard sprayer.

model mid-1960s Allman range included the 40 gallon Speedispray 40, the Six/Seventy and the Fifteen/Seventy with 21 ft wide booms and 70 gallon galvanised tanks. The Allman Fieldspray, Fieldspray Major and Spraymaster high/medium and low volume sprayers had 110 gallon polythene tanks and self-fillers.

Optional Vibrajet spray nozzles, developed by Plant Protection, were fitted to Allman sprayers and some other makes in the mid-1960s. The no-drift Vibrajet nozzle with a 6 psi working pressure had an oscillating sleeve driven by a small 12 volt electric motor. Each Vibrajet had a 72 in spraying width when used at the recommended height of 30 in above the crop.

The late 1960s Allman sprayer range included the trailed low-volume Model 200 and the mounted Contractor sprayer suitable for a wide range of chemicals and liquid fertilisers. The 210 gallon Model 200 with a five section 33 ft or optional 39 ft boom had a galvanised steel tank. The specification for the Allman Contractor with application rates of up to 60 gallons per acre at 4 mph included a 110 gallon plastic tank, polypropylene pipe work and low pressure steel or brass fan nozzles.

10.14. Accessories for the late 1950s Vigzol Blitzweed sprayer included special booms for hops, soft fruit and vineyards.

10.15 The Allman Spraymaster with a 110 gallon polyethylene tank, 33 ft spray boom and roller vane pump cost £242 10s when it was launched in 1966.

10.16 This 70 gallon Allman Six/Seventy sprayer had seven Vibrajets spaced at 30 in apart on the 21 ft spray boom.

Allman mounted and trailed sprayers for 1971 included the low volume Model 55 with an 18 ft spray boom and the 100 gallon Fieldspray with a 24 ft boom and application rates of up to 40 gallons per acre. Trailed models included the 360 with a 360 gallon Alkathene tank and a 33 ft boom suitable for spray chemicals and liquid fertilisers as well as the 60 gallons per acre Model 200 with a 10 or 12 m boom.

Simplex Dairy Equipment at Cambridge entered the sprayer market in the mid-1960s when they introduced the Simplex 50 and 75 mounted sprayers. The 50 gallon Simplex 50 low with an 18 ft boom and a gear pump applied up to 24 gallons per acre at 4 mph while the Model 75 with a 24 ft spraying width and a roller vane pump had an output of up to 45 gallons per acre.

Mid-1960s Evers & Wall crop sprayers included the 100 gallon Evrall 100 with a 24 or 30 ft spray boom which was advertised as being eminently suitable for spraying liquid fertilisers. The Evrall 100 and the 40 gallon Hardi Evrall, with a 21 ft boom and application rate of up to 25 gallons per acre, had plastic tanks and non-corrosive plumbing systems.

The early 1960s Hart Senior and Junior sprayers made by the Hart Agricultural Engineering Co were suitable for most popular makes of tractors. The 6–50 gallons per acre Hart Senior with a 60 gallon tank and the 40 gallon Hart Junior with an application rate of 5–40 gallons per acre had 21 ft spray bars. When equipped with a special pump, they could be used to spray dilute acids. The Hart Senior sprayer

10.17 The two-section spray boom on the Allman Model 55 folded vertically for transport.

10.18 *In 1965 the Croptex 220 gallon trailed sprayer with hydraulic folding and height control for the 40 ft spray boom cost £665.*

cost £60 in 1963 and the Junior was listed at £40.

Croptex trailed and mounted sprayers with polythene tanks were made by Crop Protection Ltd at Grantham in the 1960s and 1970s. The top-of-the-range Croptex 330 trailed sprayer had a 33 ft or optional 40 ft spray bar, a 330 gallon tank and a roller vane pump. Hydraulic rams to fold the spray bars and adjust the working height were optional extras.

Sales literature for the Croptex 220 gallon trailed sprayer with a polythene tank, 40 ft spray bar, high volume pump and self-filler suggested that it was 'probably the finest sprayer in the world'. The 90 and 130 gallon mounted Croptex sprayers had roller vane pumps with 32 and 40 ft spray bars respectively.

High-density polythene tanks, plastic spray bars and nylon reinforced pipe work were used in the mid-1960s to give Dorman trailed and mounted sprayers maximum protection from corrosive chemicals. Mounted models started from the 40 gallon Utility low volume sprayer with a gear pump and 18 ft plastic spray boom priced at £61. At the top end was the 100 gallon Super high-volume sprayer which, with a roller vane pump and 31 ft 6 in plastic boom, cost £190.

The mounted 100 gallon Fenlander low/medium/high volume sprayer with a plastic tank and 31 ft 6 in boom and application rates of 10–70 gallons per acre was equally suitable for chemicals and liquid fertilisers.

Dorman sprayers for copper sulphate, sulphuric acid and similar corrosive materials were available on special order. The Dorman Sprayer Co also made 200 gallon trailed and saddle tank sprayers in the mid-1960s. The trailed model had a steel tank and a 31 ft 6 in boom and an alternative plastic tank was made for spraying liquid fertilisers. The saddle tank sprayer had two side-mounted 100 gallon tanks and a hydraulically folded 40 ft boom.

The Dorman Lightweight, Fenland and Super mounted sprayers, together with the Compact and Super trailed models, were current in 1977. The Lightweight Junior sprayer with a 200 or 300 litre tank and a 5.5 m boom was ideal for small farms and the 400 or 550 litre Lightweight Senior with a 7.3 or 8.3 m spray boom was made for the medium-size mixed farm.

The Super range with tank capacities of up to 800 litres and 8–10.7 m spray widths was aimed at the farmer contractor. The 1,200 litre trailed Compact and 1,600 or 2,000 litre Super trailed sprayers, which

10.19 *The Regu-Flo control system, which ensured a consistently accurate application rate regardless of tractor speed within a given gear was optional for Dorman Super sprayers.*

sales literature explained were designed for extensive farms where water had to be carried a long distance, had spraying widths ranging from 10 to 15.3 m. When Ransomes acquired the Dorman Sprayer Co in 1978 Dorman sprayers were made at Ely until 1980 when production was moved to Ipswich.

Improved Ransomes models of FR Junior, Standard and Senior Cropguard sprayers with bigger tanks appeared in 1961 and the FR Cropguard 100 with a 20 or 32 ft boom superseded the Senior in 1964. The new FR Cropsaver with a 40 ft spray bar mounted on the tractor front-end loader and 100 gallon rear-mounted tank was launched in the same year.

By modern standards the Cropsaver boom, made by Ransomes under licence from Cleanacres Ltd at Cheltenham, broke every rule in the farm safety inspector's handbook. Even worse, Ransomes sales literature suggested that the tractor driver would be

10.20 *The 1,200 litre Dorman Compact sprayer had separate suction fillers for chemicals and water.*

at no more risk from spray drift from the Cropsaver than he would from a rear-mounted sprayer.

The 150 gallon Cropguard 150 and Cropsaver 150 with a 48 or 60 ft spray boom replaced the earlier models in 1966. The 60 ft Cropsaver had claimed work rates of up to 20 acres an hour. Sales literature explained that the new boom design eliminated bounce and the feeler legs at the outer ends of the boom would keep it level across its full width. Launched in 1968, the 350 gallon Cropguard 350 with a three-section 32 ft boom and adjustable wheel track settings was Ransomes' first trailed sprayer. The Cropguard tank and pump was also meant for the Cropsaver front-boom sprayer.

10.21 Dorman made sprayer kits with a 100 gallon plastic tank for the Land Rover. The roller vane pump was driven by a small petrol engine.

A new range of Cropguard 85 and 105 sprayers appeared at the 1969 Smithfield Show. A 25 ft 6in boom was used on the Standard Cropguard while the Super Cropguard had a 31 ft 6 in spraying width. Cropsaver improvements in 1970 included a 105 or 155 gallon polyethylene tank and a diaphragm pump. Following Ransomes' acquisition of the Dorman Sprayer Co in 1978 Dorman sprayers were made at Ely and later at Ipswich. The Dorman name was dropped in 1981.

Ransomes' ageing sprayer models were improved and renamed in 1978 as the Cropguard 400, 500 and 700 and the trailed Cropguard 350 became the 1600. In the early 1980s the Ransomes and Dorman sprayer range included four Cropguard models, two Ransomes-built Dorman trailed models and mounted Dorman Merlin, Falcon, Hawk and Eagle models with polyethylene or glass fibre tanks.

Time was running out for the Ransomes farm machinery division but the firm introduced the Micronair 700 and the prototype CTV 2500 self-propelled crop treatment vehicle before Electrolux

10.22 The first Ransomes Cropsaver sprayers with a 40 ft boom had an output of up to 15 acres in an hour.

10.23 The Ransomes Cropguard 700 replaced the ageing Cropguard 155 in 1978.

bought its farm machinery business in 1987. Features of the pivot-steer CTV 2500 included an 80 hp Ford industrial diesel engine, a seven-section 80 ft gimbal-mounted spray boom controlled with hydraulic rams, hydrostatic drive to all four wheels and a 550 gallon glass-reinforced plastic tank.

Several new names including Berthoud, Evrard, Lely and Technoma joined the more familiar ones seen on crop sprayers in the late 1970s. Features of Lely Import mounted and trailed sprayers included polyethylene tanks, stainless-steel plumbing and quick-fit non-drip plastic nozzles. The mounted 88

10.24 The Ransomes Hawk 700 sprayer with a 150 gallon tank had a 40 ft boom with quick-fit ceramic nozzles.

gallon Conquest with a 23 ft boom was the most basic model in the Lely sprayer range. Trailed models with 40 ft wide spraying widths had 250–400 gallon tanks and an optional automatic electric spraying pressure control system.

The Lely Hydraspin controlled droplet sprayer, launched at the 1982 Smithfield Show, applied up to 36 gallons per acre with its ten hydraulically driven rotary atomisers mounted on a 40 ft boom.

Lely entered the self-propelled sprayer market in the late 1980s when it introduced the Hi-Trac 125 and 250 with a 1,600 cc Volkswagen diesel engine and a demountable 1,600 litre tank and a 12–24 m spray boom. The Hi-Trac could also be used with a twin-disc 16 cwt Lely Centerliner fertiliser broadcaster.

The 1984 Allman sprayer catalogue included fifteen different mounted machines and three trailed sprayers, the most expensive being the 500 gallon Model 2300L with a 40 ft spraying width which cost £5,590.

10.25 The hydraulically driven atomiser spray heads on the 40 ft wide Ransomes Micronaire 700 air-assisted sprayer boom applied atomised chemical at a rate of between 2.8 and 28 gallons per acre.

10.26 The Lely front-mounted tank was used to mix or transfer chemical to the rear tank while on the move.

There were eight versions of the mounted Allman Unibuilt sprayer with 50–220 gallon plastic tanks and 20–39 ft spray booms.

The Allman Farmer range of sprayers 'designed with the grassland farmer in mind' with low-density chemical-resistant polyethylene tanks appeared in the late 1980s. The 50 gallon Farmer sprayer had a three section 19 ft 6 in wide boom and the 130 gallon Farmer 600 had a 39 ft wide five-section boom. Other late 1980s Allman models included the Allman Contractor range of mounted and trailed sprayers, improved Unibuilt models and Ace mounted sprayers.

Self-propelled sprayers, originally developed in the late 1960s, came into widespread use, especially on large arable farms in the early 1980s. Hubert Sands, a Norfolk farmer and contractor, made one of the first self-propelled sprayers which he built round a David Brown tractor in 1967. It was used for the Sands contracting business and in 1976 Sands Agricultural Machinery was established to manufacture SAM sprayers.

The 1981 Sprayers in Action event was a launchpad for self-propelled sprayers with an 80 ft Armer Salmon skid unit on a Mercedes MB Trac, an Allaeys Forward control sprayer and the latest model of SAM among the exhibits. Other mid-1980s sprayers included the self-propelled Atkinson Lightfoot, the Chaviot and the Mann's Gazelle low ground-pressure sprayers.

10.27 The Lely Hi-Trac 250 self-propelled sprayer had a five-speed gearbox with a top speed of 18 mph.

10.28 The late 1980s Allman Unibuilt 825 sprayer could have a hydraulic or manually folded 39 ft wide boom.

10.29 *Optional extras for the Allman Model 1525L trailed sprayer included a chemical mixer tank, a clean water container and a foam bout marker.*

The Chaviot with flotation tyres and a 45 hp Volkswagen engine was typical of a new breed of self-propelled low ground-pressure sprayers with work rates of up to 200 acres per day.

The tracklaying low ground-pressure Gazelle with a 120 hp Ford diesel engine and independent hydrostatic drive to each track had a 330 gallon Tecnoma sprayer with the option of a 40 or 80 ft wide spray boom.

Controlled droplet spraying with a Tecnoma Girojet, Ulvamast and a few other makes was popular in the mid-1980s. Application rates as low as 2 gallons

10.30 *The Allman CTV 2500 crop treatment vehicle, was developed by Ransomes but it was still only at prototype stage when Ransomes closed its farm machinery division.*

per acre were possible with a set of Tecnoma Girojets spaced across a 12 m spray boom. The speed of the 145 mm diameter spinning discs, driven by 12 volt electric motors at a speed of between 1,100 and 4,200 rpm, was adjusted with a rheostat. The French-built Girojet, which received chemical at the centre of the disc, sprayed mist-like droplets by gravitational force in a 140 degree flat fan pattern.

The Ulvamast controlled droplet sprayer was different from the others with a single rotary atomiser on a vertical arm attached to the chemical tank. The makers explained that the Ulvamast gave total crop protection at a work rate of 50 acres an hour using

10.31 *The late 1970s Sands forward control sprayer had a 350 gallon tank, hydrostatic steering and an air-conditioned cab.*

10.32 *Tecnoma Girojet spinning disc atomisers were driven by 12 volt electric motors.*

10.33 The Degania sleeve boom sprayer had a PVC air sleeve above the spray bar.

10.34 The Allman Airtec low volume sprayer had twin fluid nozzles.

99 per cent less water and an 80 per cent saving in chemical costs when compared with ordinary crop spraying.

The Degania sleeve boom sprayer introduced by Ferrag in 1988 was a new concept in crop spraying technology. Designed in Israel, the sleeve boom sprayer had large PVC sleeves above the spray bar. An air blast from a large pto-driven fan was used to blow chemical from conventional nozzles down into the crop. The sleeve boom sprayer, which achieved full penetration of the chemical even in dense foliage, also reduced the quantity of water used and minimised drift.

At least twenty different makes of self-propelled sprayer and sprayer skid units were on the market in the early 1990s. They ranged from the lightweight Frazier Agribuggy and similar machines with a 1,500 litre tank and a 24 m spray boom for the larger farm to the SAM and other contractors' sprayers with 3,600 litre tanks and spray booms up to 24 m wide. A typical trailed sprayer had a 2,500 litre tank with 12–24 m spraying widths while the more expensive models had some form of self-steering drawbar to make the sprayer wheels follow the tractor wheel marks.

Special nozzles, which used compressed air in the process of atomising pesticides, were a feature of the early 1990s Allman Airtec twin fluid low volume sprayer. Spaced at 20 in intervals on a 39 ft wide boom the nozzles applied between 2½ and 13 gallons of atomised chemical per acre. The atomised droplets, containing up to 50 per cent air, were carried with minimum drift by an air stream on to the target. At application rates of about 6 gallons per acre the 176 gallon Airtec sprayer had work rates of 250–300 acres in a day.

Band Sprayers

Specialist spraying equipment, including band sprayers, granular applicators, chemical hoes and weed wipers, were developed over the years to meet particular farming needs.

Weed control in sugar beet was a mechanical task, either by hand hoeing or with a horse- or tractor-drawn inter-row hoe and if it rained within a few

10.35 Band spraying with a Vicon sprayer mounted on a Monodrill seeder unit toolbar.

10.36 Ransomes band sprayers had a 30 gallon tank and a set of factory-matched nozzles.

hours of hoeing many of the weeds survived. The development of pre-emergence herbicides for sugar beet in the early 1960s resulted in the application of these new chemicals when the crop was drilled. Overall spraying was expensive but applying a narrow band of chemical to the freshly drilled rows with a band sprayer was a cheaper option.

Band sprayers made by Dorman, Ransomes and Vicon and mounted on the seeder unit toolbar were usually successful in killing weeds that germinated in the rows and a steerage hoe dealt with weed seedlings between the rows. The reign of the band sprayer was short lived as within a few years the chemical companies introduced selective herbicides for the sugar beet crop and the band sprayers were left in the barn.

Some band sprayer tanks were mounted on the front of the tractor but it was more usual for the tank to be carried on the drill toolbar. A pto-driven pump supplied chemical from the tank via a pressure control unit to a set of factory-matched nozzles mounted above the rear press wheel on each seeder unit.

Early Dorman band sprayers had a 30 gallon front-mounted tank, a 3½ gallon per minute gear pump direct-mounted on the pto shaft and a set of no-drift nozzles which applied a 7 in wide band of chemical over each row. Later 66 or 88 gallon Dorman band sprayers were used with seeder units drilling up to 15 rows at the same time. With the optional 24 or 32 ft wide spray boom for the Vicon four-, five- or six-row band sprayer could also be used for overall spraying.

Ransomes made band sprayers with non-drift hollow cone nozzles for the popular makes of four- to ten-row seeder units. The tank was either mounted in front of the tractor or on the seeder unit toolbar. The last Ransomes band sprayers were made in 1970.

Tractor-mounted weed wipers, including the Vicon Wedgewik, Tecnoma Top-Weeder and Matrot Mobilcord, were used to touch-treat tall weeds in growing crops with a herbicide-impregnated rope or wick. The Wedgewik and Top-Weeder had a stationary rope impregnated with chemical from a tractor-mounted tank. The Mobilcord had continuously moving ropes on a system of pulleys which were saturated with herbicide as they passed through an immersion tank.

10.37 The herbicide-impregnated rope on the Vicon Wedgewik applicator delivered a fatal dose of herbicide to tall weeds growing above the crop.

Index

Below is just a small selection of the wide range of agricultural books and DVDs which we publish.

For more information or a free illustrated catalogue please contact:

Old Pond Publishing Ltd
Dencora Business Centre,
36 White House Road, Ipswich,
Suffolk IP1 5LT United Kingdom
Website: www.oldpond.com
Tel: 01473 238200

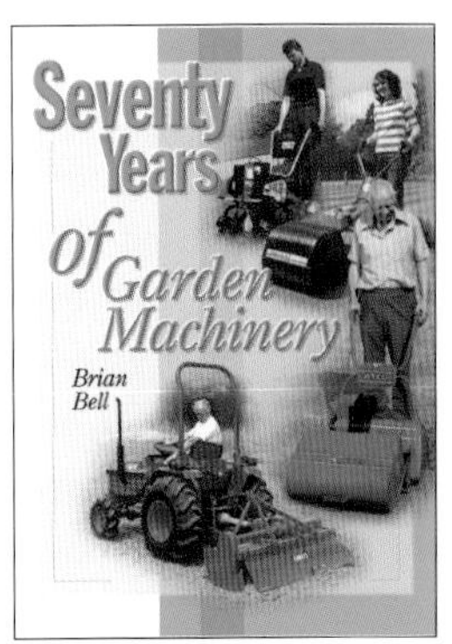

Seventy Years of Garden Machinery

Brian Bell

Covering the period 1920-1990, Brian Bell's book deals with the development of smaller machines: 2-wheeled garden tractors, rotary cultivators, 4-wheeled ride-on tractors, ploughs, drills, cultivators, sprayers, grass-cutting equipment, small trucks, miscellaneous estate items. Hardback book.

Ransomes, Sims and Jefferies

Brian Bell

Ransomes invented the self-sharpening ploughshare and made, among other products, steam engines, lawn mowers, trolleybuses, threshers, reach trucks, tractors, subsoilers, disc harrows, sprayers, mowers, root-crop equipment and machinery for export. Brian Bell's comprehensive book emphasises 20th century farm machinery. Hardback book.

Farm and Workshop Welding

Andrew Pearce

Andrew Pearce's detailed and highly illustrated book covers MMA, Mig/Mag, gas welding and cutting, TIG, plasma cutting, cast iron, pipe welding, hardfacing, soldering, welding plastics, drills, taps and dies, basic blacksmithing. Hardback book.

Tractor Restoration: paintwork

Alan Davies

Professional restorer Alan Davies shows how he sets about preparing and spraying a tractor. He includes beating out damage, filling, stopping, spraying panels with primer and colour. He shows specific techniques for wheels and the chassis. The demonstrations are clear and detailed. DVD

Farm Machinery Film Records 1, 2 & 3

Brian Bell

This footage from the 1940s and '50s shows a wide range of farm machinery which was in development at the time. It comes from the archive of the National Institute of Agricultural Engineering and its Scottish counterpart. Altogether, over seventy machines are included in the programmes. Experimental prototypes from the NIAE itself are shown. DVDs

About the Author

Brian Bell MBE

A Norfolk farmer's son, Brian played a key role in developing agricultural education in Suffolk from the 1950s onwards. For many years he was vice-principal of the Otley Agricultural College having previously headed the agricultural engineering section. He established the annual 'Power in Action' demonstrations in which the latest farm machinery is put through its paces and he campaigned vigorously for improved farm safety, serving for many years on the Suffolk Farm Safety Committee. He is secretary of the Suffolk Farm Machinery Club. In 1993 he retired from Otley College and was created a Member of the Order of the British Empire for his services to agriculture. He is past secretary and chairman of the East Anglian branch of the Institution of Agricultural Engineers.

Brian's writing career began in 1963 with the publication of *Farm Machinery* in Cassell's 'Farm Books' series. In 1979 Farming Press published a new *Farm Machinery*, which is now in its fifth enlarged edition, with more than 25,000 copies sold. Brian's involvement with videos began in 1995 when he compiled and scripted *Classic Farm Machinery Vol 1.*

Brian Bell writes on machinery past and present for several specialist magazines. He lives in Suffolk with his wife Ivy. They have three sons.

Books and DVDs by Brian Bell

Books

Farm Machinery 5th Edition
Fifty Years of Farm Tractors
Machinery for Horticulture (with Stewart Cousins)
Ransomes, Sims and Jefferies
Seventy Years of Farm Machinery: 1. Seedtime
Seventy Years of Garden Machinery
Tractor Ploughing Manual

DVDs

Acres of Change
Classic Combines
Classic Farm Machinery Vol. 1 1940-1970
Classic Farm Machinery Vol. 2 1970-1995
Classic Tractors
Farm Machinery Film Records Vol 1 Grain Grass and Silage
Farm Machinery Film Records Vol 2 Autumn Work and Rootcrops
Farm Machinery Film Records Vol 3 Testing and Prototypes
Harvest from Sickle to Satellite
Ploughs and Ploughing Techniques
Power of the Past
Reversible and Conventional Match Ploughing Skills
Steam at Strumpshaw
Thatcher's Harvest
Tracks Across the Field
Vintage Match Ploughing Skills
Vintage Garden Tractors